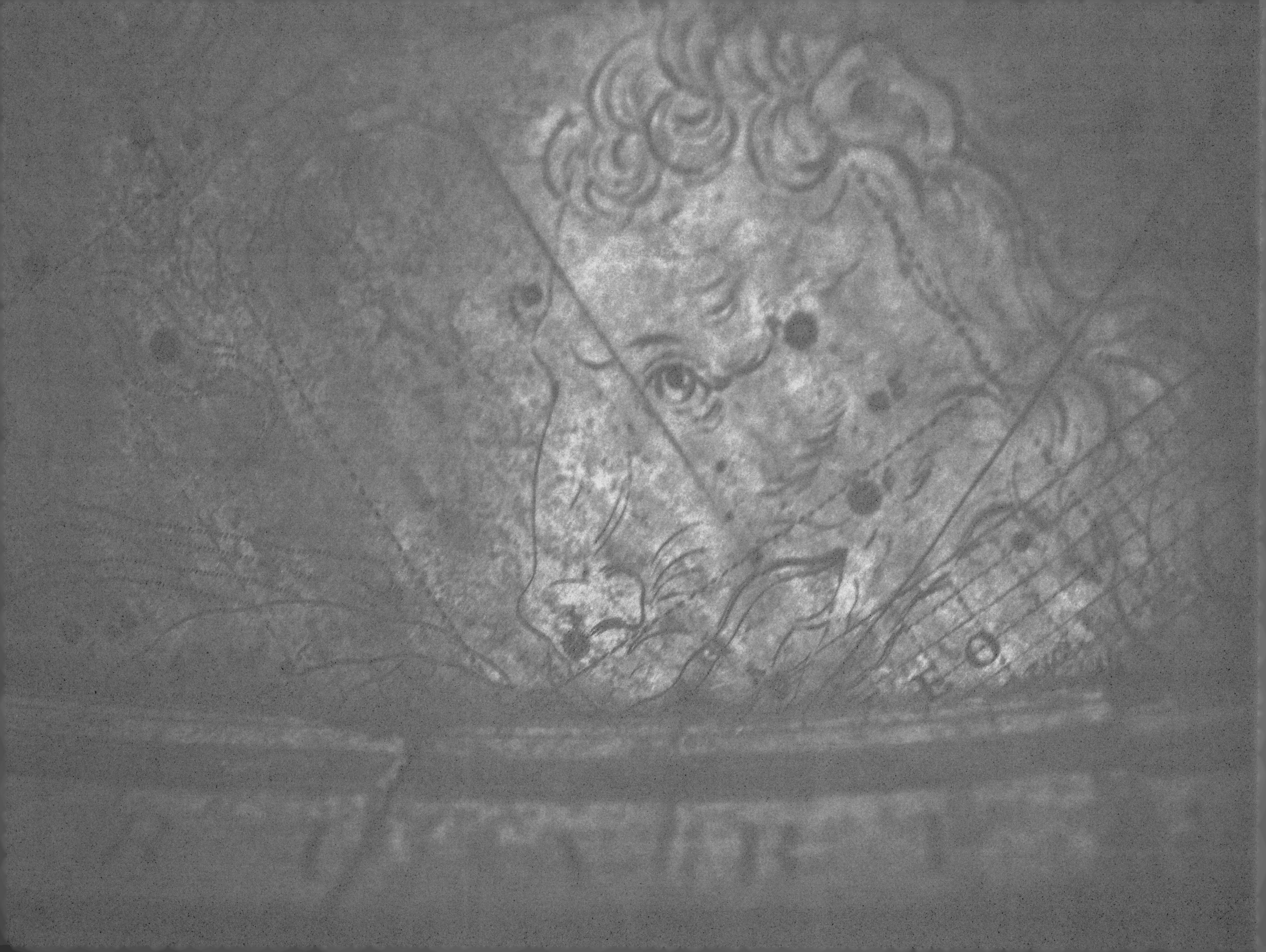

EVA KOLEVA TIMOTHY

LOST IN LEARNING

THE ART OF DISCOVERY

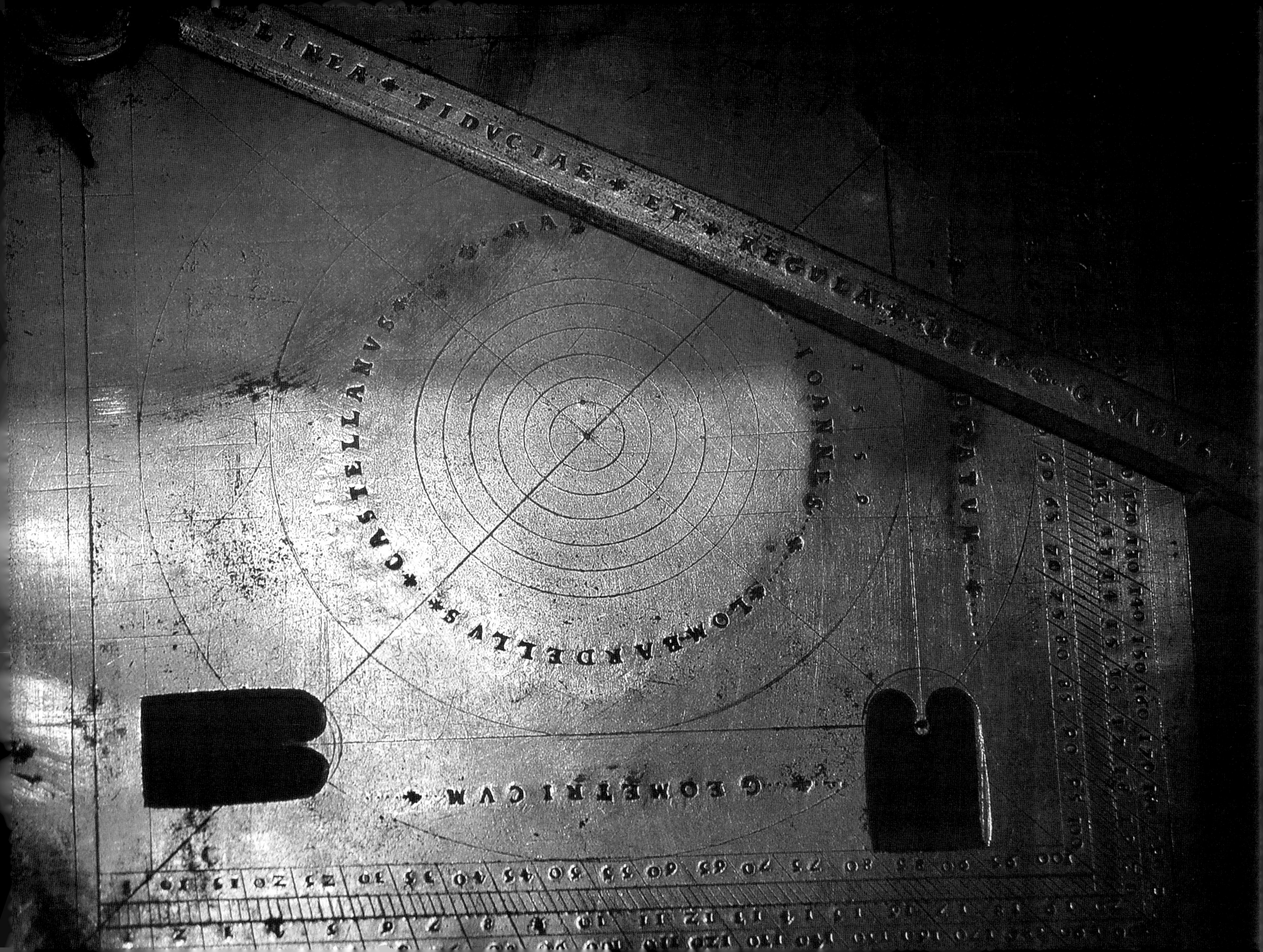
LINEA FIDVCIAE ET REGVLA
IOANNES
CASTELLANVS
LOMBARDELLVS
1559
GEOMETRICVM

EVA KOLEVA TIMOTHY

LOST IN LEARNING

THE ART OF DISCOVERY

FOREWORD BY RALPH WINDLE
ESSAY BY ADAM TIMOTHY

ATHENAEUM PUBLISHING

Dedicated To

Ian, London & Skye

Whose young lives have taught me the art of true discovery

CONTENTS

"Youth of delight come hither
And see the opening morn,
Image of truth new born"

William Blake (1757 – 1827)
Songs of Innocence

FOREWORD

It says much for the haunting significance of Eva Timothy's new work that it brought William Blake's "Songs of Innocence and Experience" straight back to mind — and that is no mean company in which to move!

It's not just the special medium — juxtaposing eye-popping images with challenging words; though it's worth noting that Blake, printmaker as well as poet and visionary, was a "craftsman" first, tending his artefacts through to full fruition and, no doubt, with inky fingers; not unlike this present day photo-visionary in her working studio. Both share great creative integrity in that alone.

Yet the truer analogy is in the mission. In times darker than our own, Blake exposes through his art whatever he feels narrows our perception or stifles the human spirit. His visions were not just escapist fantasies but insights into the everyday reality we could have if it were re-created by freshly-awakened and re-energised people. Art, he thinks, is for changing things; has the potential power to move us in newer, better directions.

And Eva Timothy thinks so, too, on the evidence of this remarkably inventive book. She gains, by being less strident, more congenial in her tone; but loses nothing in the persuasive strength and clarity of conviction of her message. The emotive, almost iconic, images of discovery come to the eye surprisingly refreshed by some brilliant innovations of angle, light and composition; among which the words emerge seamlessly but with an inescapable insistence on being heard.

For all its rich evocation of history, however, this monograph looks forward more than back; and its fresh look at childhood leaps far beyond the usual nostalgia for a lost innocence.

Something much more important is happening here, and it connects in an exciting, novel way with one of the mainstream developments in contemporary literature and art.

We are seeing a surge in what the critics are calling "crossover" creativity. Just as "Alice's Adventures in Wonderland", written by Lewis Carroll to amuse a niece, have always enthralled parent as much as child; so with the more recent "serious" and sustained "children's epics" — Tolkien's "Lord of the Rings", Ursula LeGuin, Philip Pullman, even "Harry Potter" — we are receiving signals of a promising re-convergence of the imaginative worlds of so-called "young" and "old". I can now read the book, take in the movie, without an eye-brow raised at all my very apparent years!

Eva Timothy's special eye has seen the deeper significance of this. It is that, "young" or "old", we are forever just one person, never two. So whatever we off-load into life's lumber-room, when the going gets tough, it must surely not be the very life-enhancing qualities of excitement, wonder, curiosity and experiment which got us out of our cradles in the first place.

To re-discover, re-explain and re-energise such truths is the great mission of art and the artist.

This is what "Lost in Learning" so consummately does.

Ralph Windle
Founder/Director of the Creative Value Network.
Writer ArtScience Commentator
Oxford, 2010

In Search of New Worlds I

From *Imago Mundi* by Pierre d'Ailly, 1410

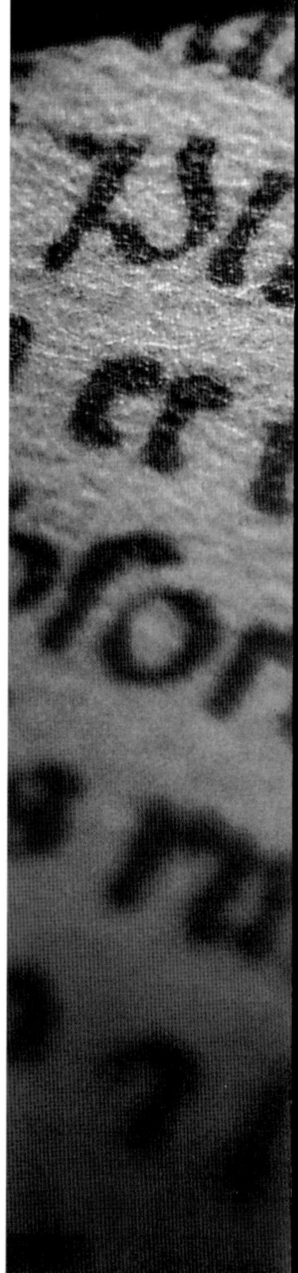

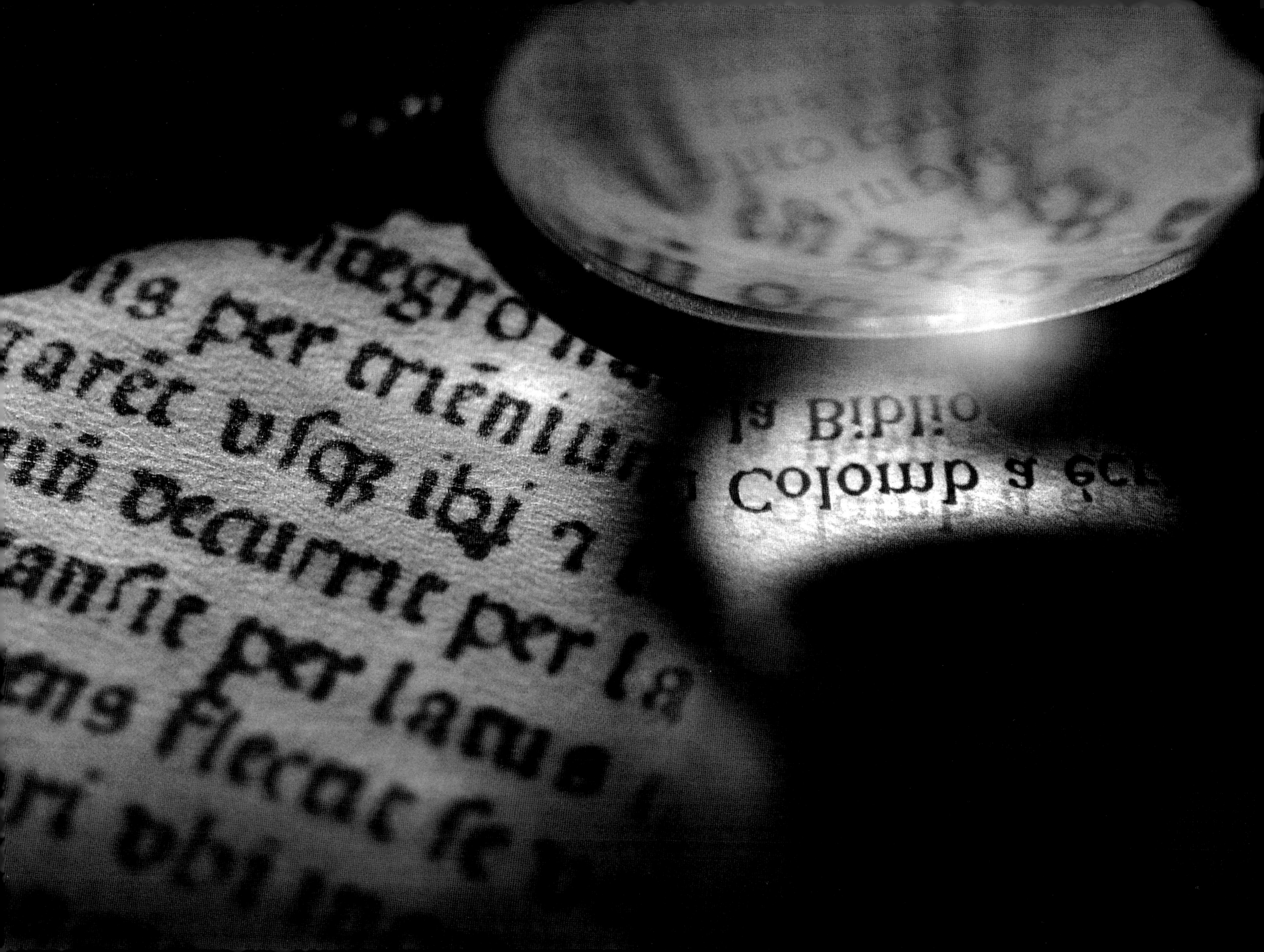

"Twenty years from now you will be more disappointed by the things that you didn't do than by the ones you did do.

So throw off the bowlines.
Sail away from the safe harbor.
Catch the trade winds in your sails.
Explore. Dream. Discover."

-Attributed to Mark Twain

Introduction

In the midst of lives jam-packed with bills to pay, errands to run, deadlines to meet, events to attend, round-the-clock news broadcasts, and click-of-a-mouse information overload, we all stand in need of quiet moments to sharpen our vision.

We are immersed in a world of hyper-media marketing, saturated with visual cues.

Yet, most of what we consume visually each day does little to fulfill or inspire us as people. In fact, it dulls us and lulls us into a world of materialism and cheap sensation.

I see art as one of humanity's great antidotes to the plague of hyper-consumption, because art has the power to help us gain perspective on those things that matter most.

Photography in particular has the capacity to shape our worldview, to influence those details upon which we choose to focus.

Outside all the dulling distractions, there is so much in life worth exploring, worth wondering about, and worth seeking after.

The Lost in Learning project explores education as a creative, exhilarating adventure filled with unknowns and great "Aha!" moments along the way.

As the mother of three, I have the opportunity daily to observe the insatiable curiosity of children and their overwhelming enthusiasm for discovery.

The creative spark I see in them is no different from the inspiration that propelled and guided some of history's greatest learners in their own courses of monumental discovery.

Immersing myself in this project and studying the lives of some of history's greatest explorers has taught me how vital our perspectives on education and learning are to the vibrancy of our lives.

"Education" can be something we plod wearily through to the end of the semester, trying only to meet the requirements or gain the favor of those who designed the course. But how much better to embrace learning as that lifelong quest through which we create meaning as we pursue our dreams!

May all your learning voyages be filled with inspiration, and may they lead you to where you long to be.

Carpe Diem—Sub Specie Aeternitatis!

Eva

The Voyage

The seed of greatness lies in each of us.

Its cultivation is the labor of a lifetime.

And yet...from the moment we are born, we inherit a gift so profound, so extraordinary, that it should never be taken for granted.

This gift is simply our ability to know or to understand something today which we did not fathom the day before.

In short, we are able to LEARN...and through learning, we open the doors to our most daring dreams.

Yet, in an age where the primary motivations for education are often pecuniary in nature, we tend to lose sight of the pure poetic elegance of learning as the substance of life itself.

True learning is INNATE and passionate. It requires neither coercion nor cajoling. It springs naturally from our desires to grow, to create, to master, and to make meaning of life.

While composed of many interim accomplishments, it is less a destination or a diploma than an epic voyage taken over the course of a lifetime.

As we embark upon that vast sea of learning, we may choose to row our way across in the straightest possible path to the other side.

Or we can set sail, propelled over the waves by the winds of our imagination and the currents of creativity.

In either case, there will be storms to weather, doldrums to deal with, and a constant realignment of our course over time.

The determining question between the two is this:

When we reach our destination, will we remain "landlubbers" or will we have become seafarers?

Will we have learned to LOVE the journey enough to continue our explorations and to set sail again and again throughout our lives? Or will a lingering sense of seasickness leave us forever landlocked?

Ultimately that will depend on what we perceive as we gaze off at the horizon.

Are there gallant dreams, noble callings, and daring visions BECKONING us in the daily pursuit of the best that is in us and around us? Or do we see just another pleasant view—something to be admired from a distance and then forgotten as we sink back into a lukewarm pool of anesthetizing escape?

As we lose ourselves in our learning, we allow ourselves to become part of something greater than self. We begin to perceive the possible though presently invisible and it is this wonderful combination of purpose and perspective which conjures the zephyrs of inspiration to bear us toward our dreams.

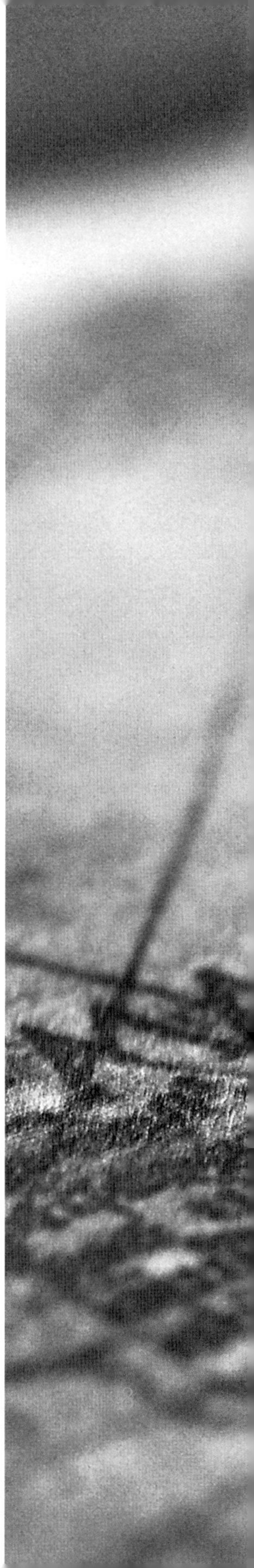

Sailing

From *Imago Mundi* by Pierre d'Ailly, 1410

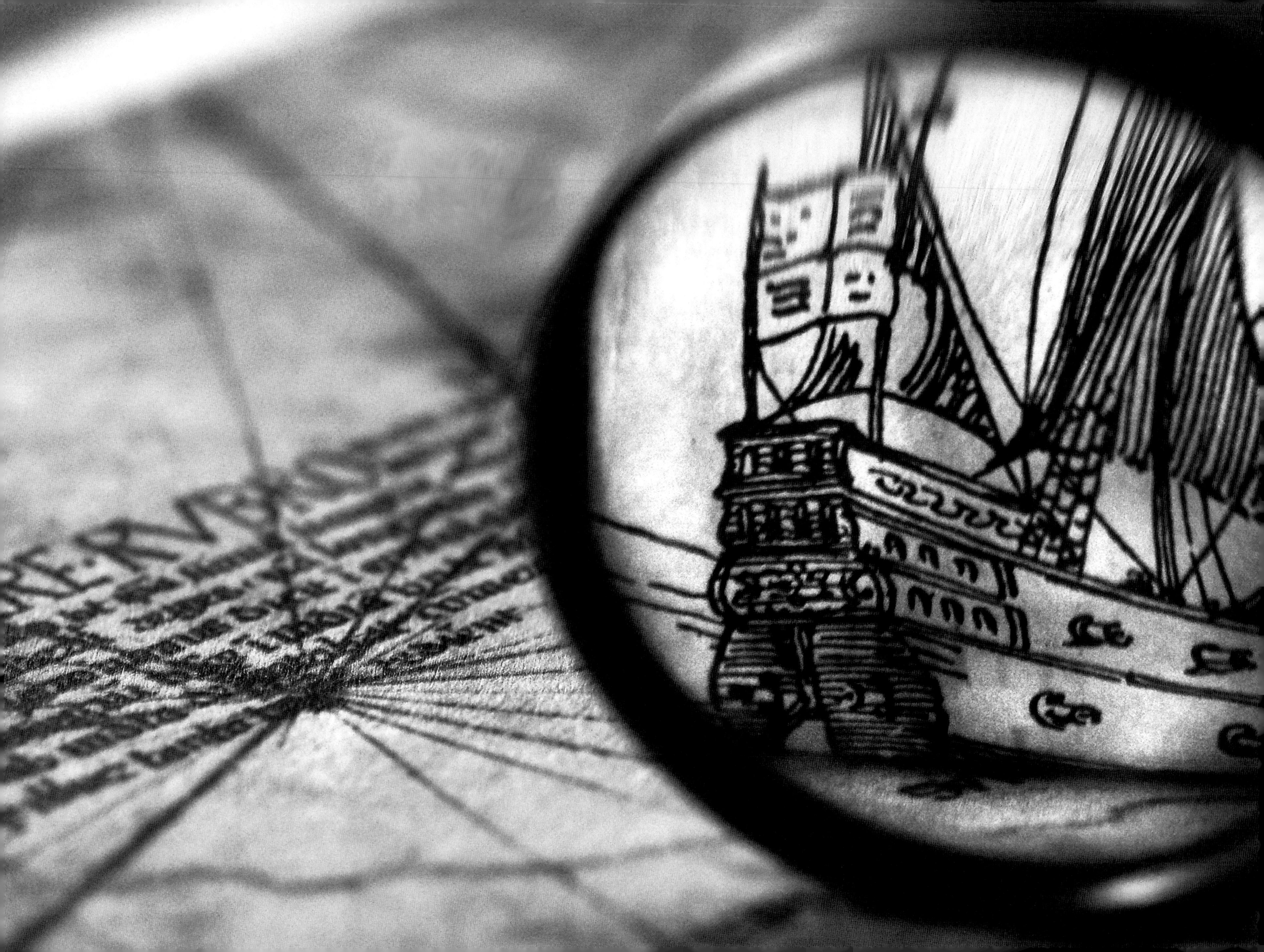

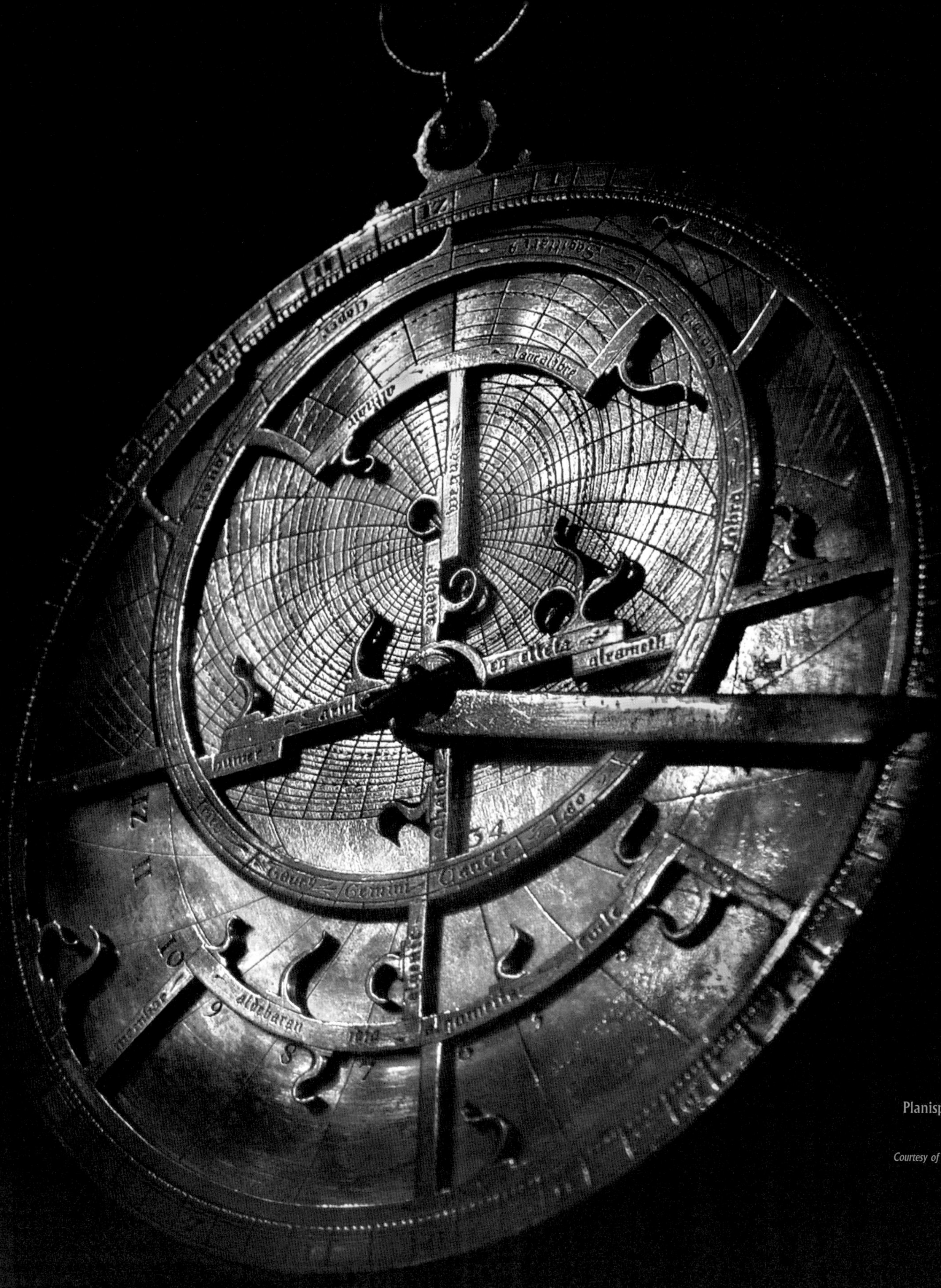

Fusoris Astrolabe

Planispheric Astrolabe, workshop of Jean Fusoris, Paris, France c. 1400

Courtesy of the History of Science Department, Collection of Historical Scientific Instruments, Harvard University

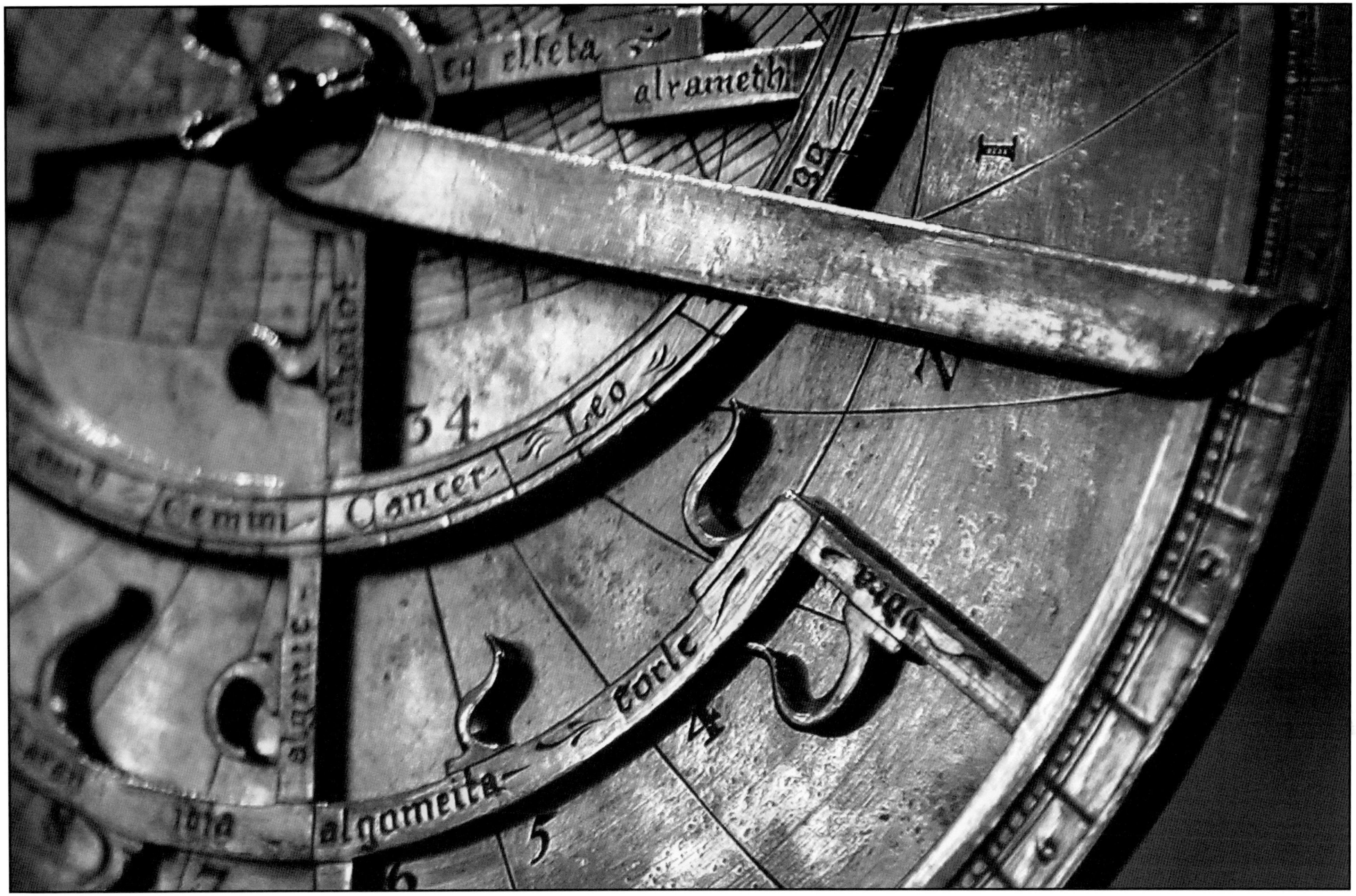
alrameth
Leo
Cancer
algameita

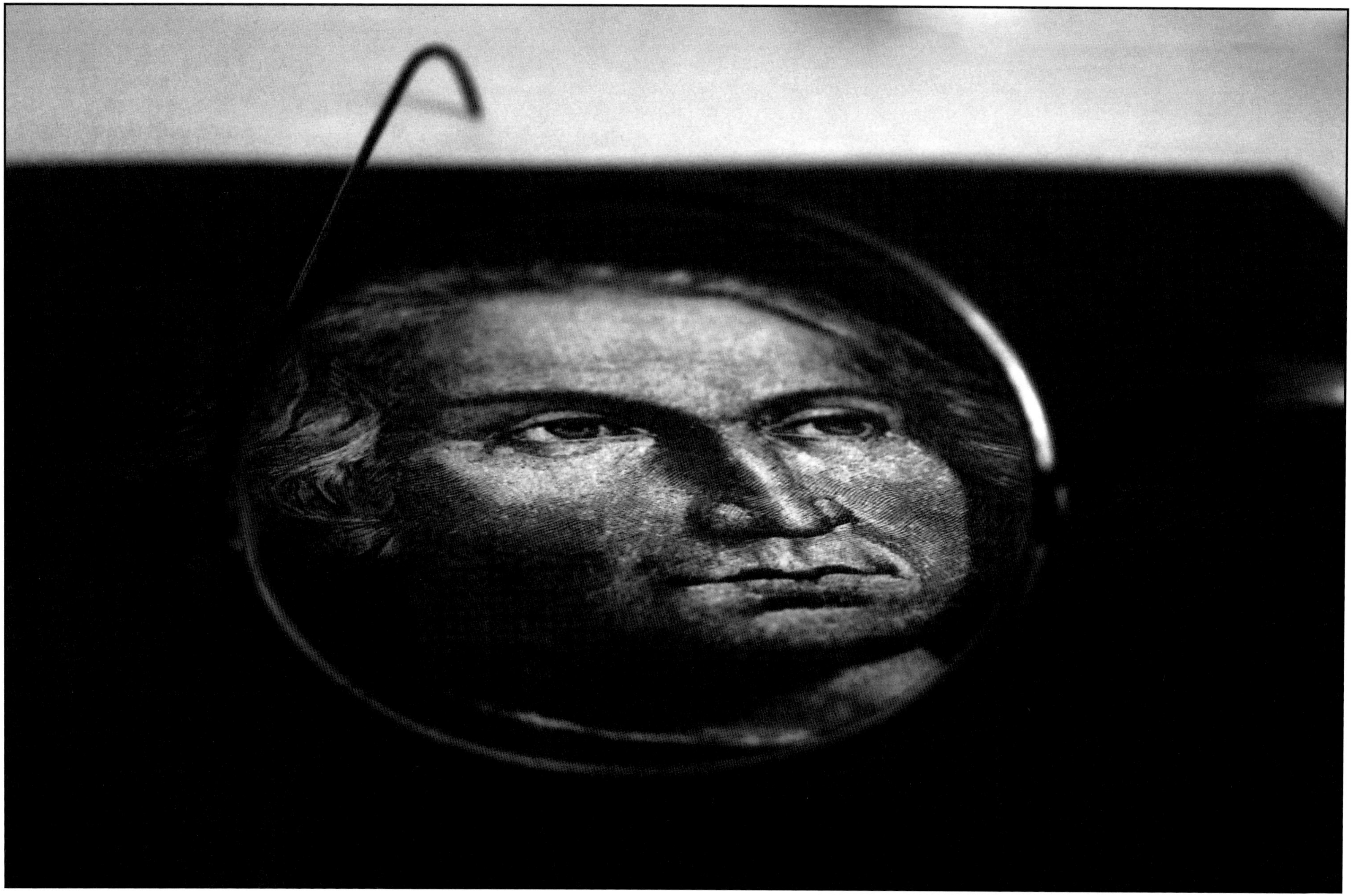

Christopher Columbus

1451-1506

Thirty days into a voyage, whose success was dubious at best, the crew aboard Columbus' ships had grown mutinous. They had embarked knowing that they might never return home; but at least, if they turned back now, there were still sufficient supplies to make the return trip.

Columbus, though, was not about to see decades of dreaming, petitioning, and laboring vanish just yet. The risk of a mutiny paled in comparison with the chance of actually discovering a westward route to the Far East.

If Columbus had been any less ambitious, he might have satisfied his curiosity within the comfort of a library. But with EXPLORATION'S FIRE running through his veins, simply ruminating over a potential sea route west would never suffice. He had to see for himself what lay westward, even if such knowledge came at the peril of life and limb.

That's not to say that Columbus didn't study. Having worked as a mapmaker and sailor, he knew all there was to know about navigation in the late fifteenth century. He was a master of the winds, and he had studied the best geographic texts available. He filled the margins of his copy of Imago Mundi by Pierre d'Ailly with his own handwritten notes as he sought to deduce the circumference of the earth and the resultant distance to the East Indies.

To prove it though, he would have to leave the known world.

And thus here he was, his supplies running low and his crew's confidence in success even lower. Yet, pressured to turn back, the captain was able to convince his crew to take just three more days of forward journeying.

On the third day, the first signs of land appeared. Shortly thereafter, the New World was sighted, and a sea path westward established. Columbus returned to Spain a hero, and the lad from Genoa was named governor of the new land.

We could say the rest is history, but it is what followed that set Columbus apart as a great explorer. Endowed with land, riches, and power sufficient for a lifetime, Columbus grew restless, and his thoughts returned to further exploration and discovery.

He had found a path west. But what lay BEYOND the land he had found? And again, what lay beyond that? That wonderful sense of curiosity impelled his return journey again and again as he laid aside all worldly comfort and put his life at stake in seeking to answer the never-ending question of what else lay westward.

In his learning Columbus had made a number of incorrect calculations. He died having no conception of what it was he had actually DISCOVERED. Yet those errors, as they played out over history, were somewhat immaterial, for he accomplished what mattered most, having lived a life of unceasing exploration.

The Admiral of the Ocean Sea

Portrait of Christopher Columbus From The Museo Naval in Madrid, Spain

The Admiral's Autograph

Facsimile of Christopher Columbus' Signature

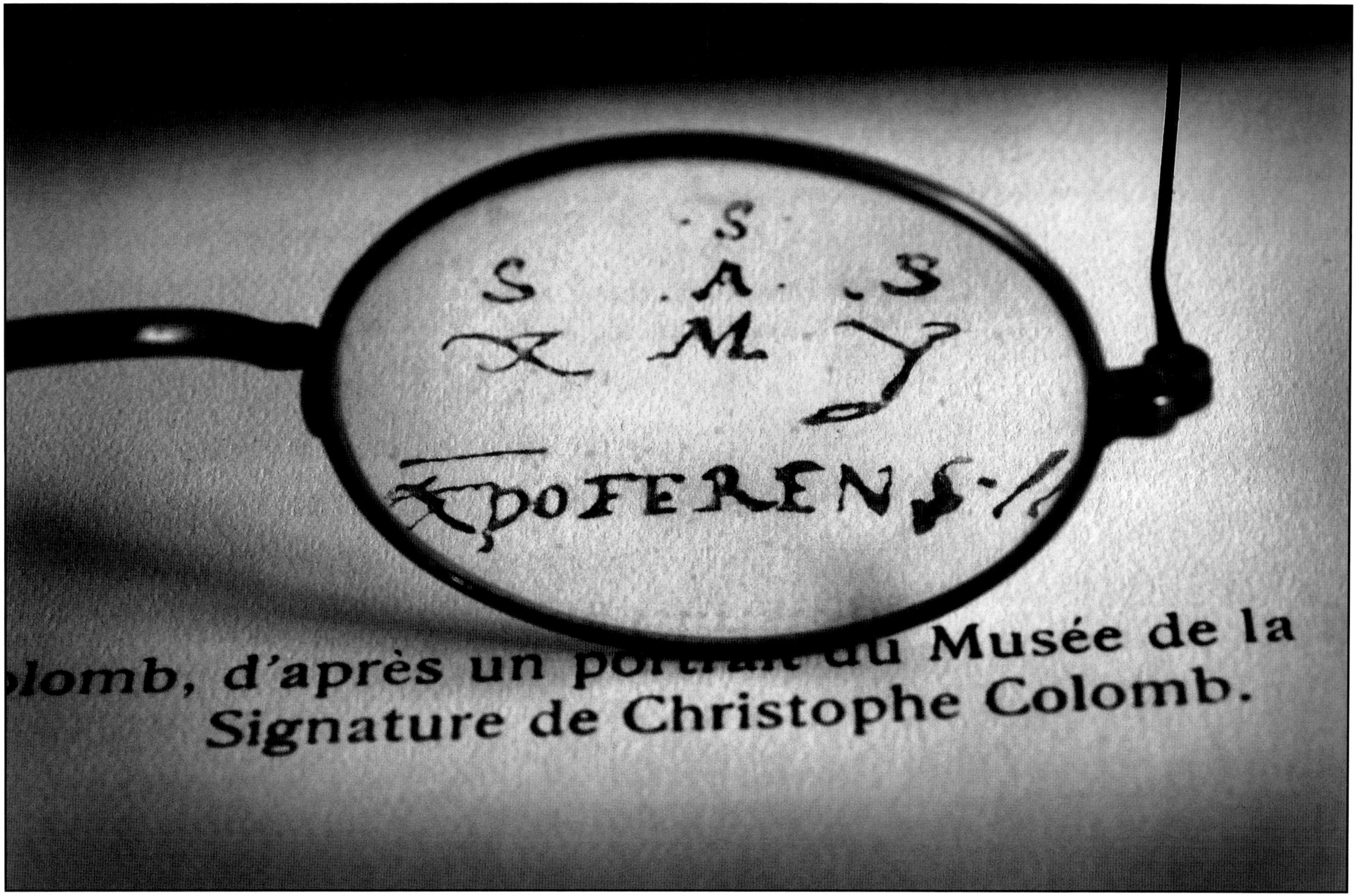

.S.
S .A. S
X M Y
Xpo FERENS
lomb, d'après un
Musée de la
Signature de Christophe Colomb.

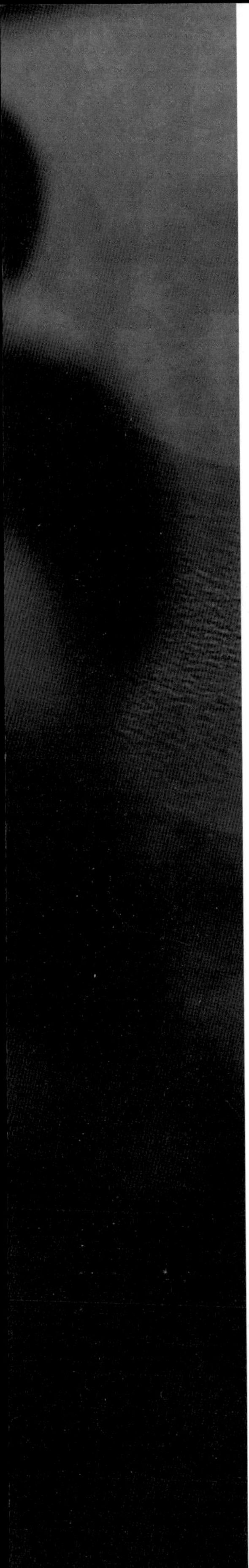

In Search of New Worlds II

From *Imago Mundi* by Pierre d'Ailly, 1410

In Search of New Worlds III

Imago Mundi by Pierre d'Ailly, 1410

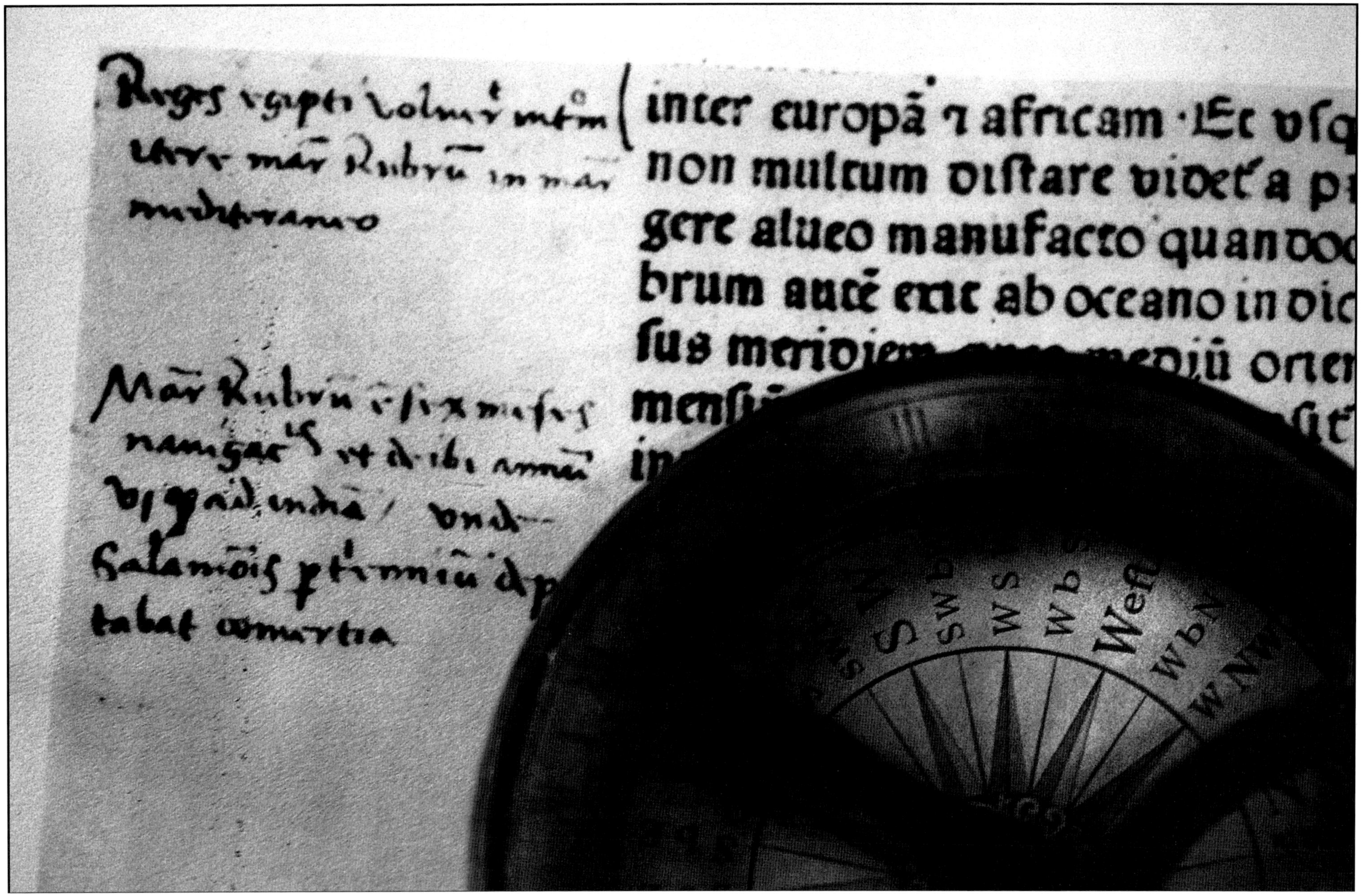
inter europā 7 africam · Et usq
non multum distare videt a p
gere alueo manufacto quando
brum autē exit ab oceano in dic
West

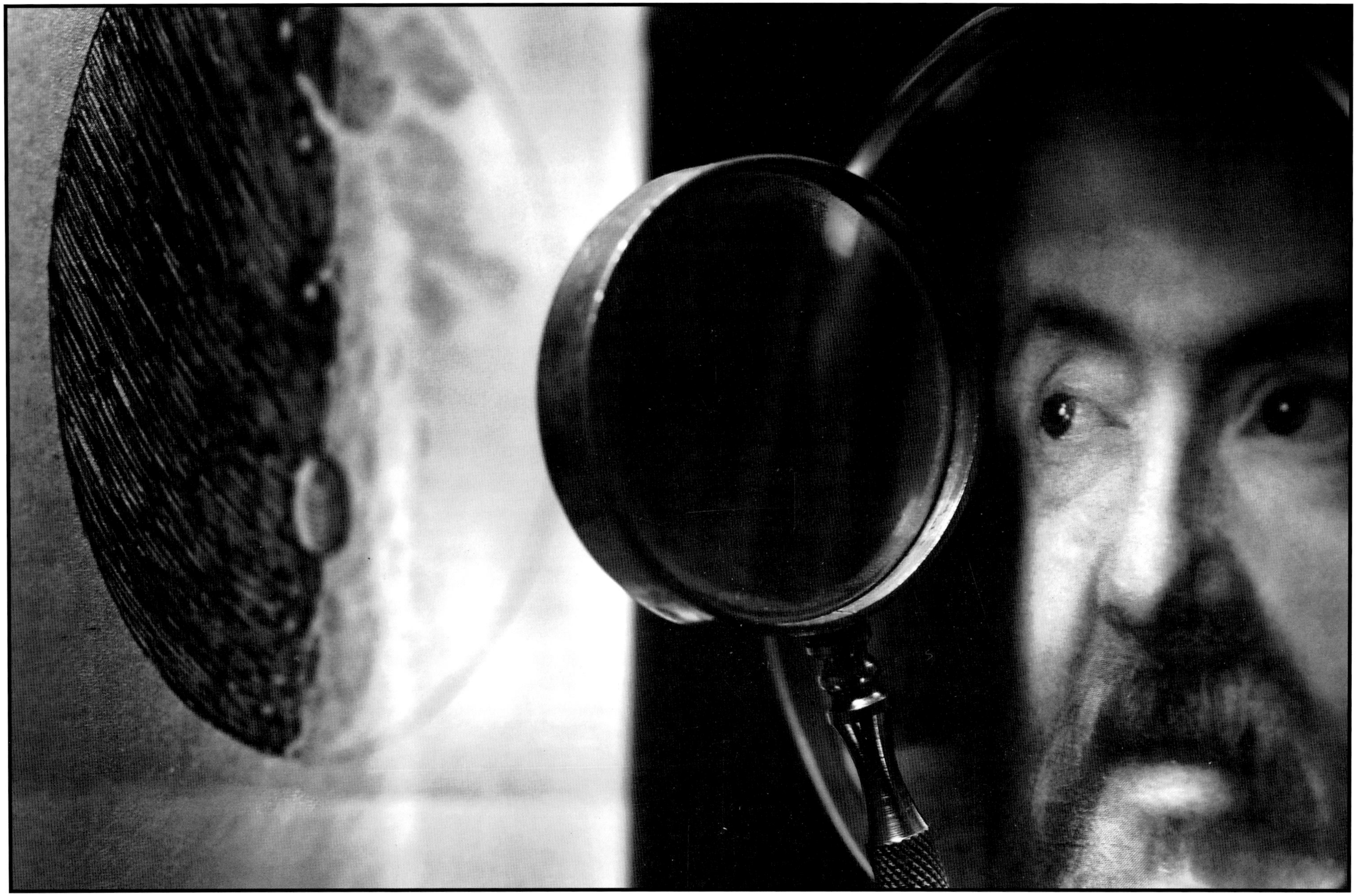

Galileo Galilei

1564-1642

For the third night straight, Galileo Galilei could not sleep. Rather, he gazed out the window, his one hand moving swiftly across the page as he sought to capture, in pencil, the miracle at the other end of the lens.

There it hung for him—so near he felt he could touch it—the silver moon with every facet revealed in glorious detail.

Galileo's telescope would animate the aristocracy of Italy with its military and commercial applications. His royalties from the invention would ensure that he and his family lived in comfort for many years to come.

Like a child with a new toy though, Galileo was most EXCITED by the implications of his invention for his own learning, as he could now survey the heavens.

In later years, Galileo would pen the words:

"I render infinite thanks to God for being so kind as to make me alone the first observer of marvels kept hidden in obscurity for all the previous centuries."

From Earth's moon he would move on to study the moons of Jupiter and develop his theory of the cosmos with the sun at its center.

Such a theory stood in stark contrast to the Aristotelian wisdom of the day, which placed the earth at the center of the universe.

In the early seventeenth century, when the only acceptable means for seeking truth was philosophizing, earthly experimentation with the celestial sphere was heresy of the highest order.

Galileo, however, stood UNDAUNTED in his determination to explore the heavens with every sense allotted him.

Indeed, Galileo was generally not one who could be intimidated into abandoning his deeply held interests.

As a young boy, his wealthy and influential father was determined that his son should become a doctor.

But finding himself unmoved by medicine, Galileo found other interests which shaped his life's learning.

Between an early fascination with music and the lute, observations of a pendulum at the local baptistery, and an introduction to Euclid's geometry, it was not long before Galileo discovered his lifelong passion for mathematics and natural physics.

While best remembered as a scientist, Galileo was, like all great learners, a seeker of TRUTH and BEAUTY wherever it was to be found. He employed his every faculty in that pursuit and refused to be silenced by the political machinations which threatened his work with censorship.

His work and his life stand as a tribute to one who paid no heed to the siren calls of comfort and ease, but who insisted on pursuing his own PASSION for learning to the journey's end.

The Astronomer

Moon Sketch from Siderius Nuncius, by Galileo Galilei, 1610

Galileo Portrait by Justus Sustermans, 1636

Galileo's Compass

Galileo's Geometrical and Military Compass

Made for presentation to the Duke of Mantua by Marc'Antonio Mazzoleni, Padua, Italy, 1604

Courtesy of the History of Science Department, Collection of Historical Scientific Instruments, Harvard University

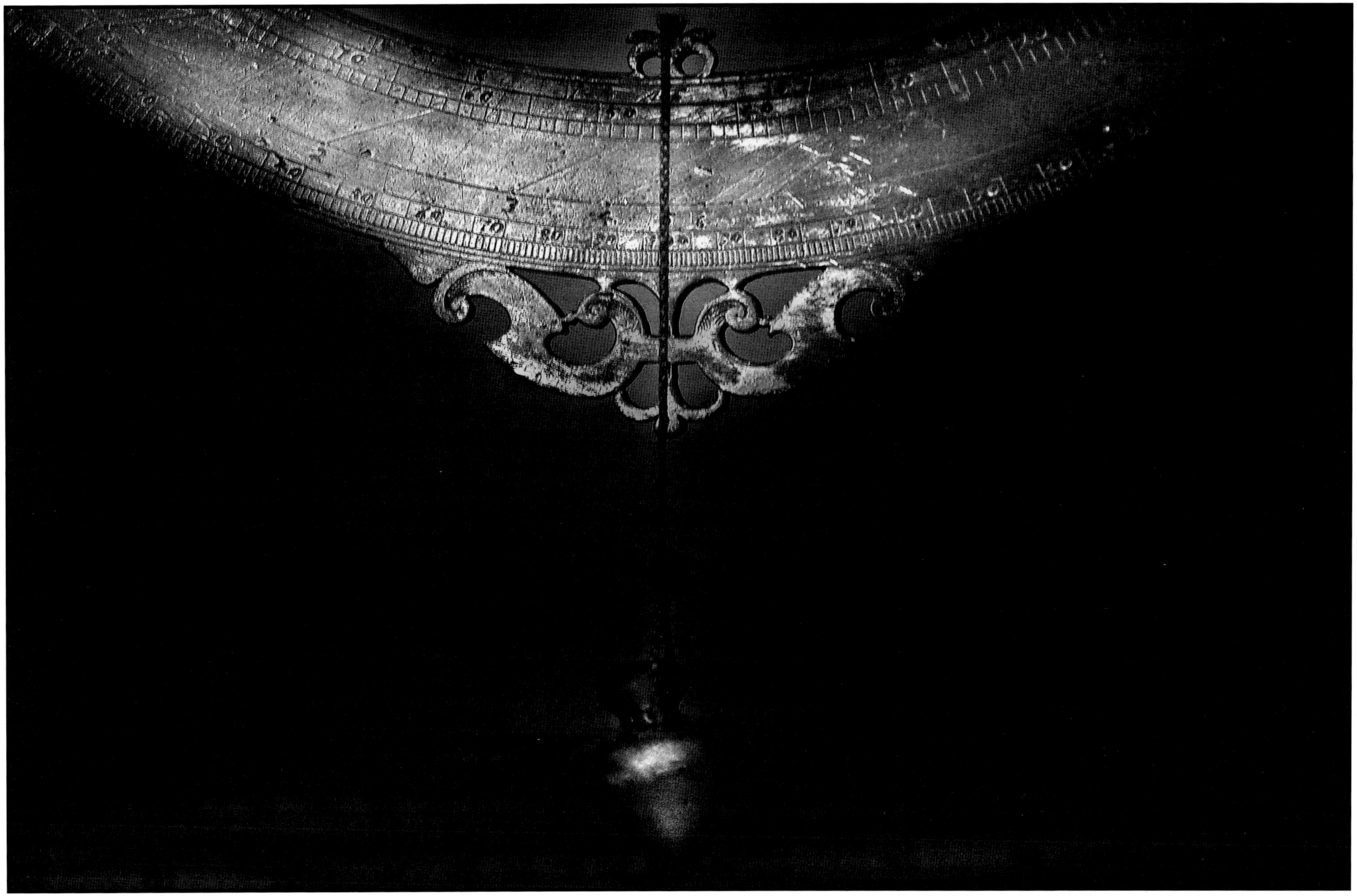

CELESTIAL SPHERE

Ptolemaic Armillary Sphere

Joseph Simon Guibot, Paris, France 1737

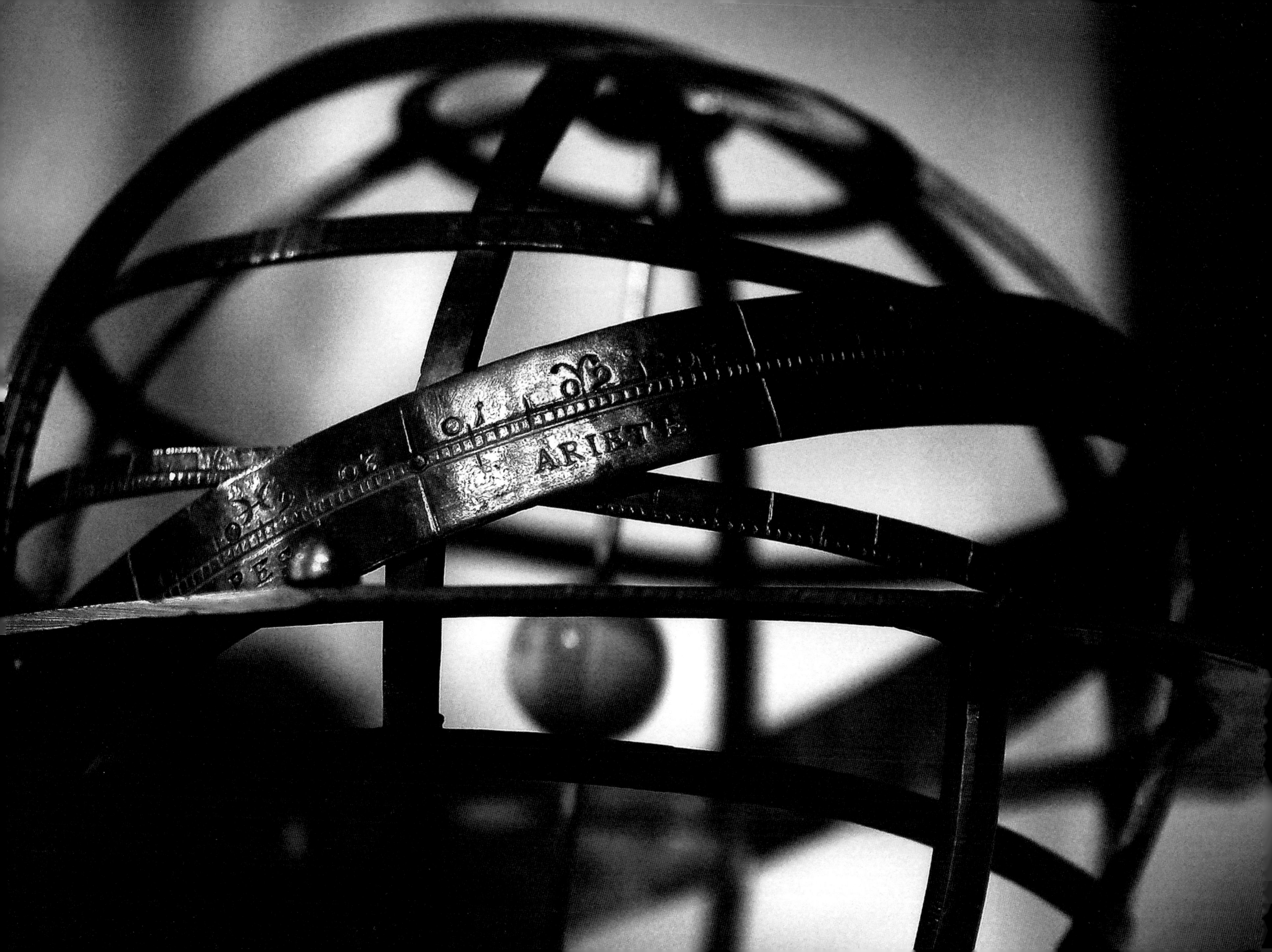
ARIETE

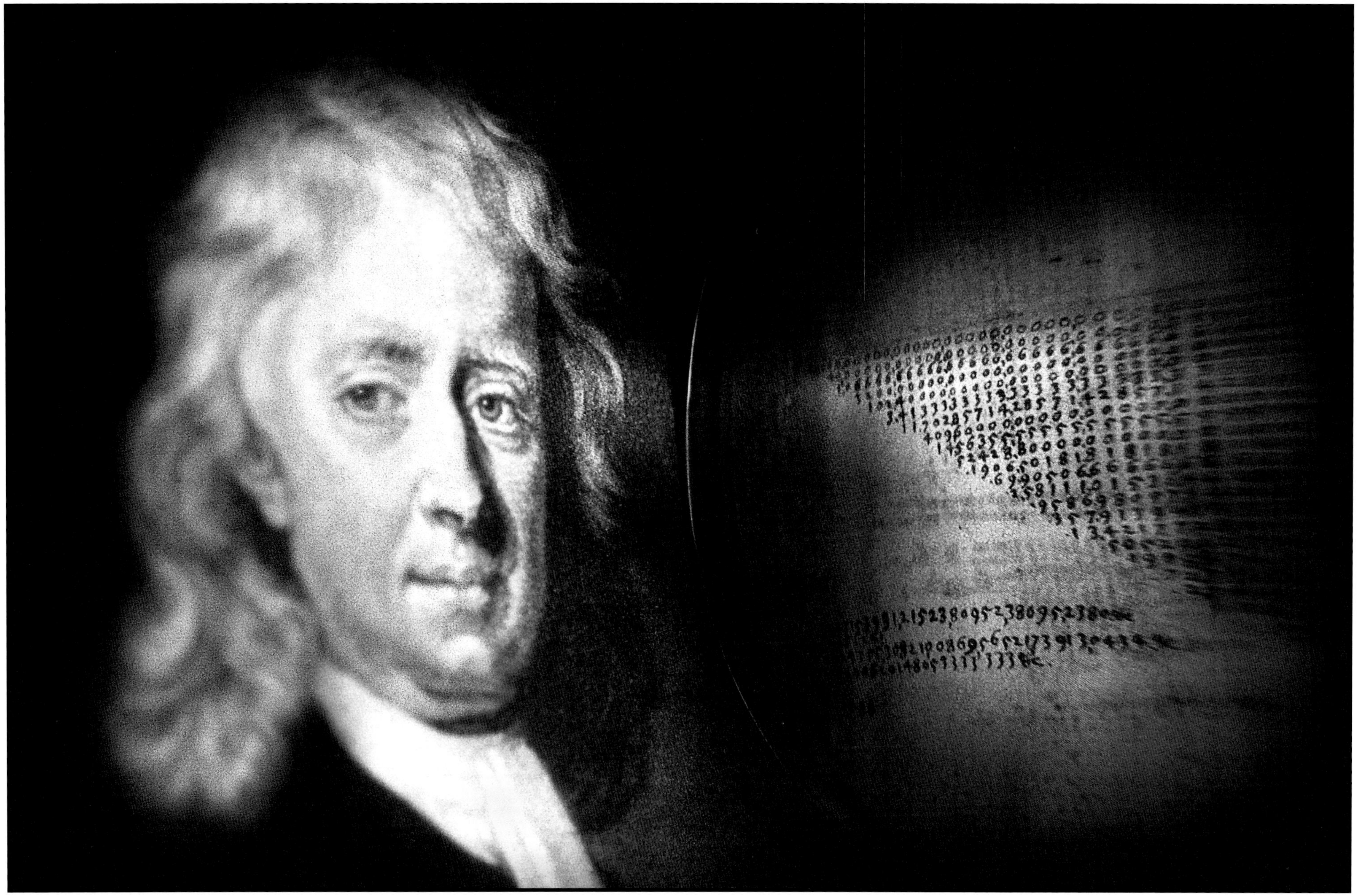

Isaac Newton

1642-1727

In the same year that Galileo completed his mortal sojourn, another bright star arrived on earth whose discoveries would redefine our view of the physical world.

The boy had an incredible eye for detail, an unceasing imagination and the TENACITY to stick with a problem until it was solved. He was often lost in reverie and oblivious to the surrounding world.

Record has it that on his way home from school, there was a large hill where all the students would dismount from their horses before climbing. Then they would remount at the summit.

More often than not, however, young Isaac Newton would forget to remount, or else forget his horse entirely, leaving it to walk home alone while he continued his daydreaming.

But his losing himself in his learning, he revealed some of the greatest insights and inspiration ever granted to man.

For the rest of the world, a falling apple was nothing more than a plummeting piece of fruit. To Isaac though, it held the KEY to the inner-workings of the Universe.

For him, there was substance in light, and there existed no natural phenomena but what could be unraveled and explained by natural law.

Through observation, experimentation, and unwavering determination, he unlocked some of the greatest mysteries of physics and stood INTREPID even before such staggering concepts as INFINITY.

Indeed, using the idea of taking infinitely small measurements, like grains of sand along the seashore, he developed his theory of calculus, which transformed mathematical science from top to bottom.

Yet for all his prestige and fame, Newton could express the modesty of a man who is deeply AWARE of how very much he still does not know.

Towards the end of his illustrious career, he penned the following:

"I know not what I appear to the world, but to myself I seem to have been only like a boy playing on the seashore, and diverting myself in now and then finding a smoother pebble or a prettier shell, whilst the great ocean of truth lay all undiscovered before me."

Portrait of Sir Isaac Newton by Enoch Sussemans

LIGHT

Title Page from *Opticks* by Sir Isaac Newton, 1704

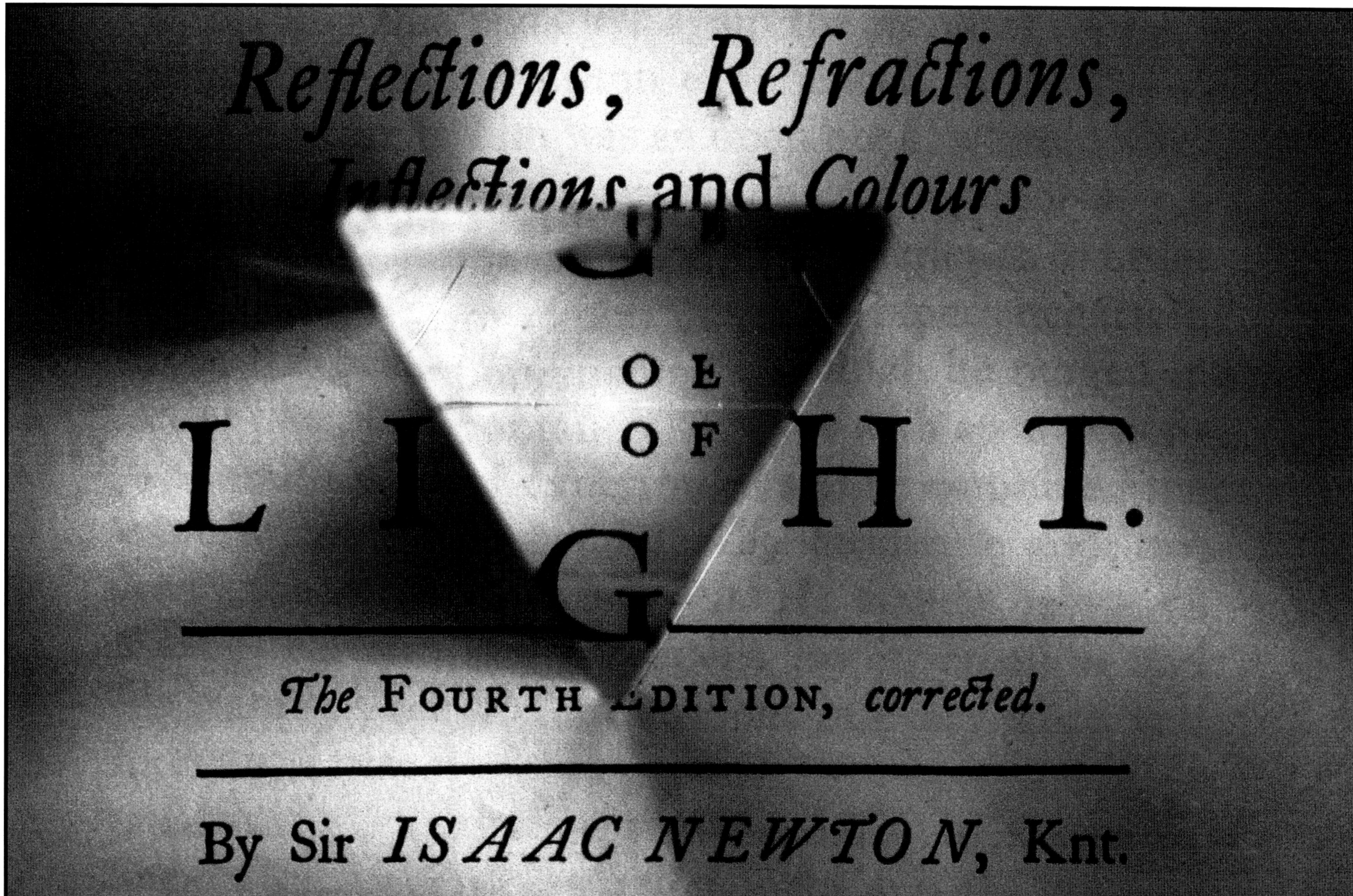
Reflections, Refractions,
and Colours
OF
LIGHT.
The FOURTH EDITION, corrected.
By Sir ISAAC NEWTON, Knt.

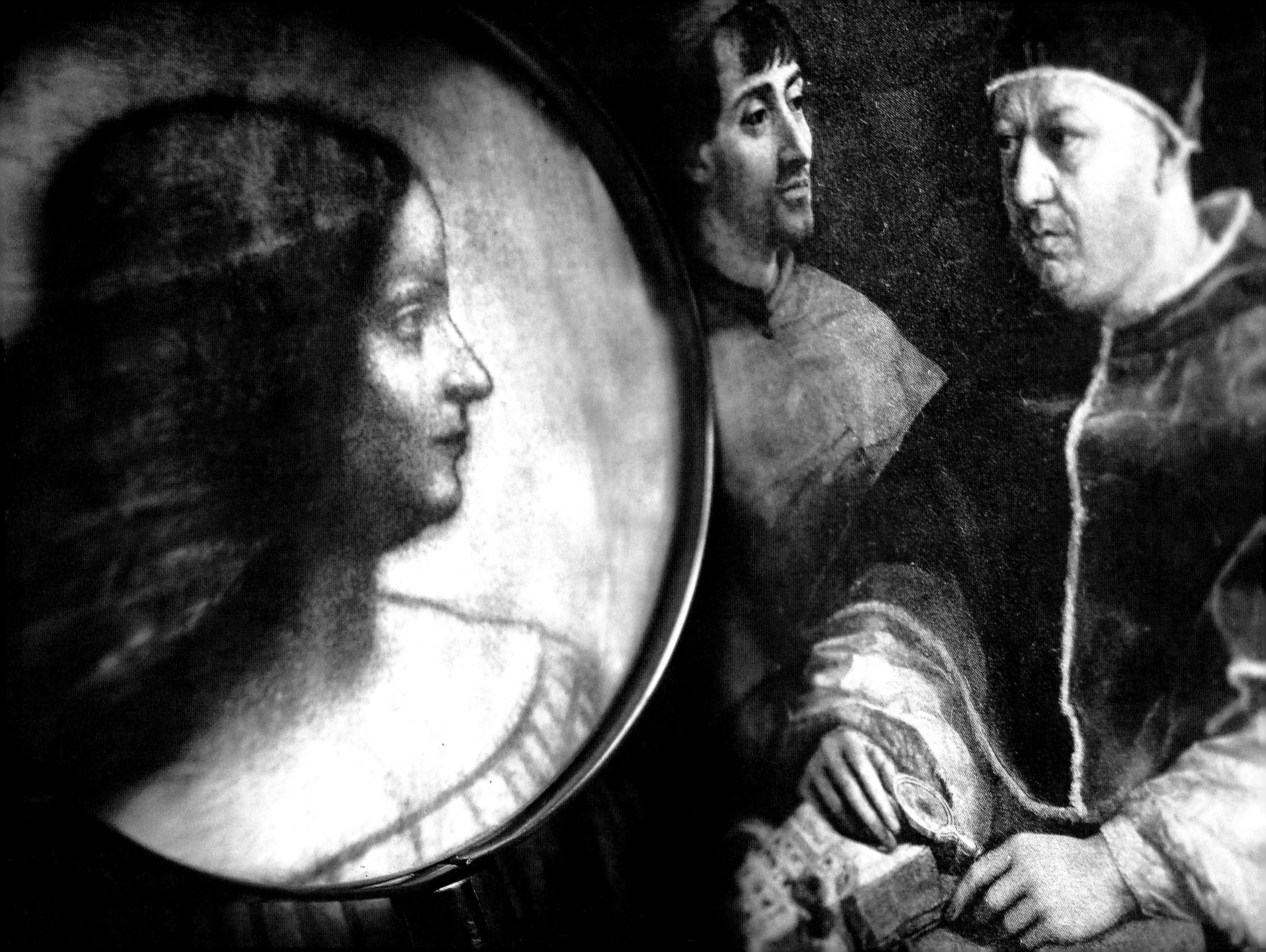

Isabella d'Este

1474-1539

While women may have been largely under-represented in the cultural works of the age, the First Lady of the Italian Renaissance, Isabella d'Este, was not a whit behind her male counterparts in her ENTHUSIASM for learning, art, and every other ennobling pursuit.

At sixteen years of age, Isabella was an accomplished musician and dancer, spoke fluent Latin and Greek, and was well versed in the politics of her day.

She was also a prolific writer, as the more than 2,000 letters she penned attest.

Born to high society and endowed with power and wealth, she might well have contented herself with a life of luxury, comfort, and pleasure seeking.

Her greatest AMBITIONS, though, were to expand knowledge and culture, and she employed virtually all of her resources on promoting the arts and education.

She established one of the first schools for young women and became one of the greatest patrons the arts have ever known.

Isabella developed a close friendship with the illustrious Pope Leo X, another great patron of education, and she sought out the greatest minds and artists to share their ideas and their talents at her court.

She was unafraid to shatter cultural glass ceilings in a male-dominated world and as the Marchesa of Mantua, she proved herself to be one of the most competent leaders of her day.

At one point she even directed the military effort to hold off invaders during a time when her husband was being held hostage in Venice.

Her can-do attitude and love for learning are an inspiration to all those determined to pursue their own AUDACIOUS dreams.

The Patrons of Learning

Isabella D'Este by Leonardo da Vinci

Pope Leo X by Raphael

The Annunciation

Detail of *The Annunciation* by Leonardo da Vinci

Uffizi Gallery, Florence, Italy c. 1472-1475

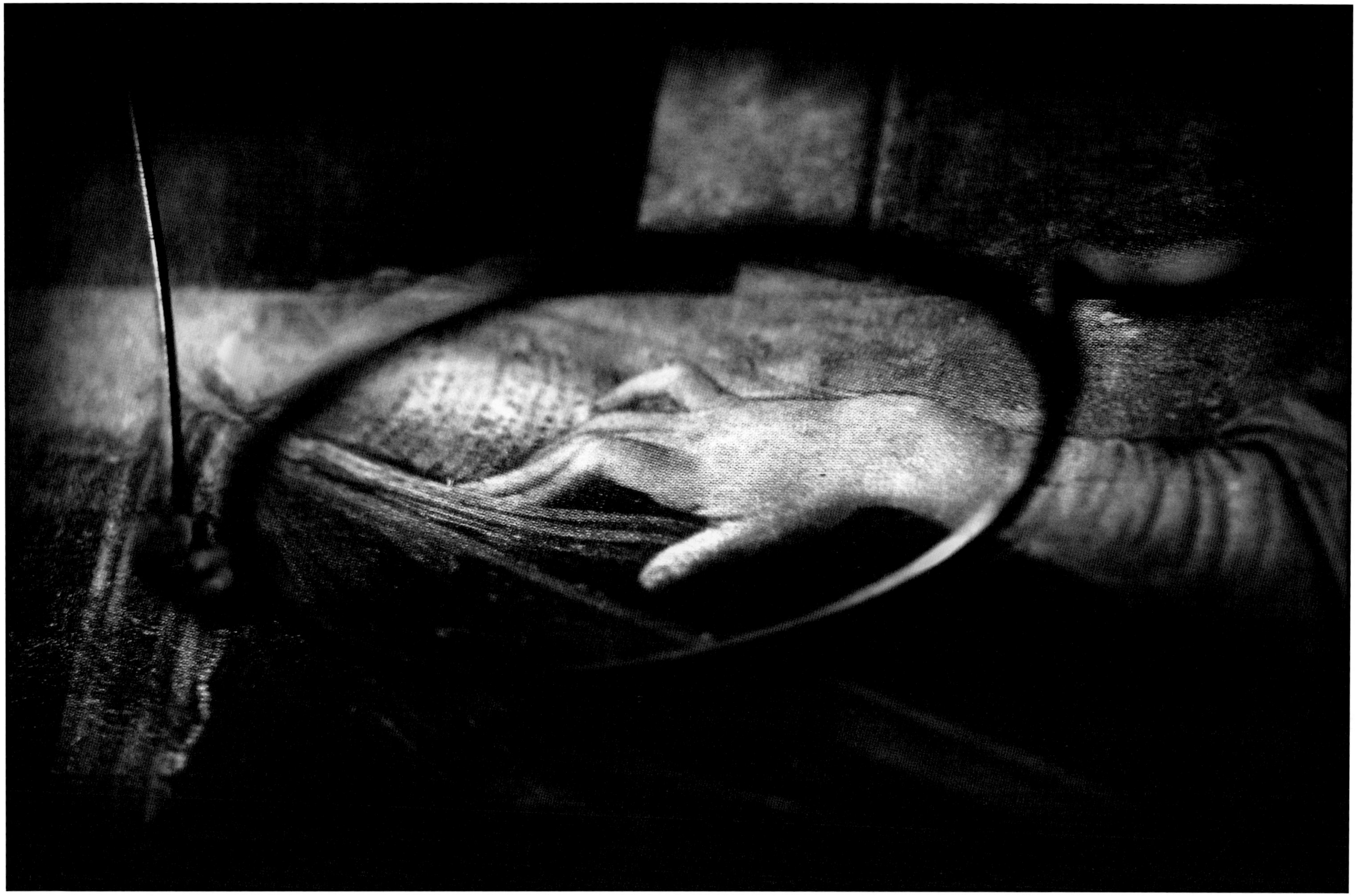

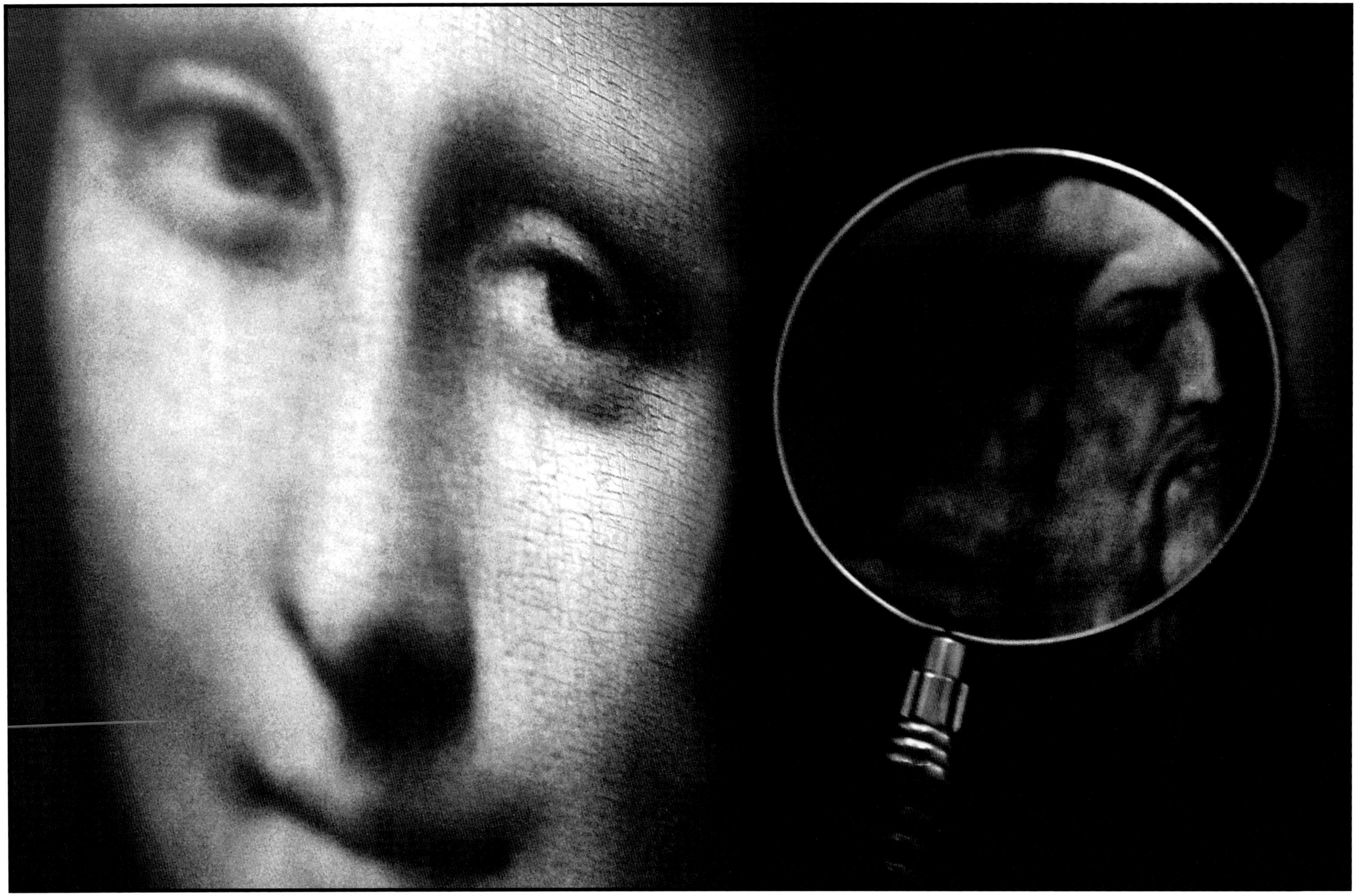

Leonardo da Vinci

1452-1519

How do we describe the man who was both the artist behind the world's most famous painting and the mind behind some of the most far-fetched and creative innovations of his time?

Leonardo da Vinci was a man whose fascination with the world and its workings knew no bounds.

Flying machines, water walking, huge pulleys, armaments, the symmetry of human form, the moon's light, cryptic codes, and the enigma of art were all things he EMBRACED and examined in great depth throughout his learning voyage.

He seemed to take pleasure in every learning experience life had to offer as he incessantly pondered and OBSERVED the world around him.

While his journals and sketches have become world renowned, he felt that all people had something worthwhile to record and hence they should keep journals of their own.

In his treatise on paintings, he wrote:

"You should often amuse yourself when you take a walk for recreation, in watching and taking note of the attitudes and actions of men as they talk and dispute or laugh or come to blows one with another...noting these down with rapid strokes in this way, in a little pocket-book which you ought always to carry with you. And let this be of tinted paper so that it may not be rubbed out...for these are not things to be rubbed out but to be preserved with the utmost diligence; for there is such an infinite number of forms and actions of things that the memory is incapable of preserving them, and therefore you should keep those sketches as your patterns and teachers."

Leonardo took in the whole world as his teacher, yet he remained his own man in all things.

When he received some of his first commissions to paint religious subjects, he was given very specific instruction as to how they were to appear; but, as was typical of his life, Leonardo paid much greater heed to his own sense of aesthetics and painted from his HEART.

The traditional halos were glaringly absent. Furthermore, the figures in his paintings appeared much more ALIVE and human than was common for such works in his day. And yet, despite all its unconventionality—or perhaps because of it—his works were considered masterpieces.

Da Vinci's breadth of talent across so many disciplines may seem extraordinary to us today. Yet for him these were all just different experiences, different lenses through which to explore, to create and to consider the boundless possibilities of learning about everything life and the surrounding world have to offer.

The Man Behind the Mona Lisa

Mona Lisa by Leonardo Da Vinci

Louvre, Paris, France 1503-1506

Blinking Moon

Sheet 2a, folio 2r

From the *Codex Leicester* by Leonardo Da Vinci, c. 1508-1512

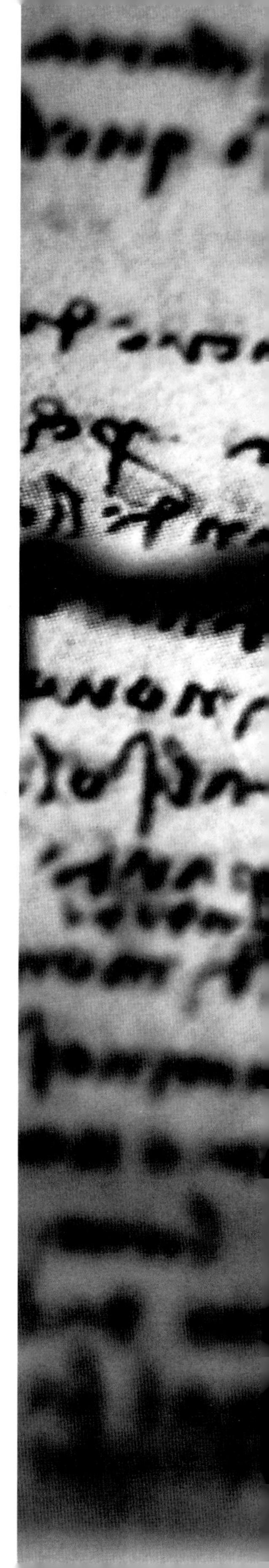

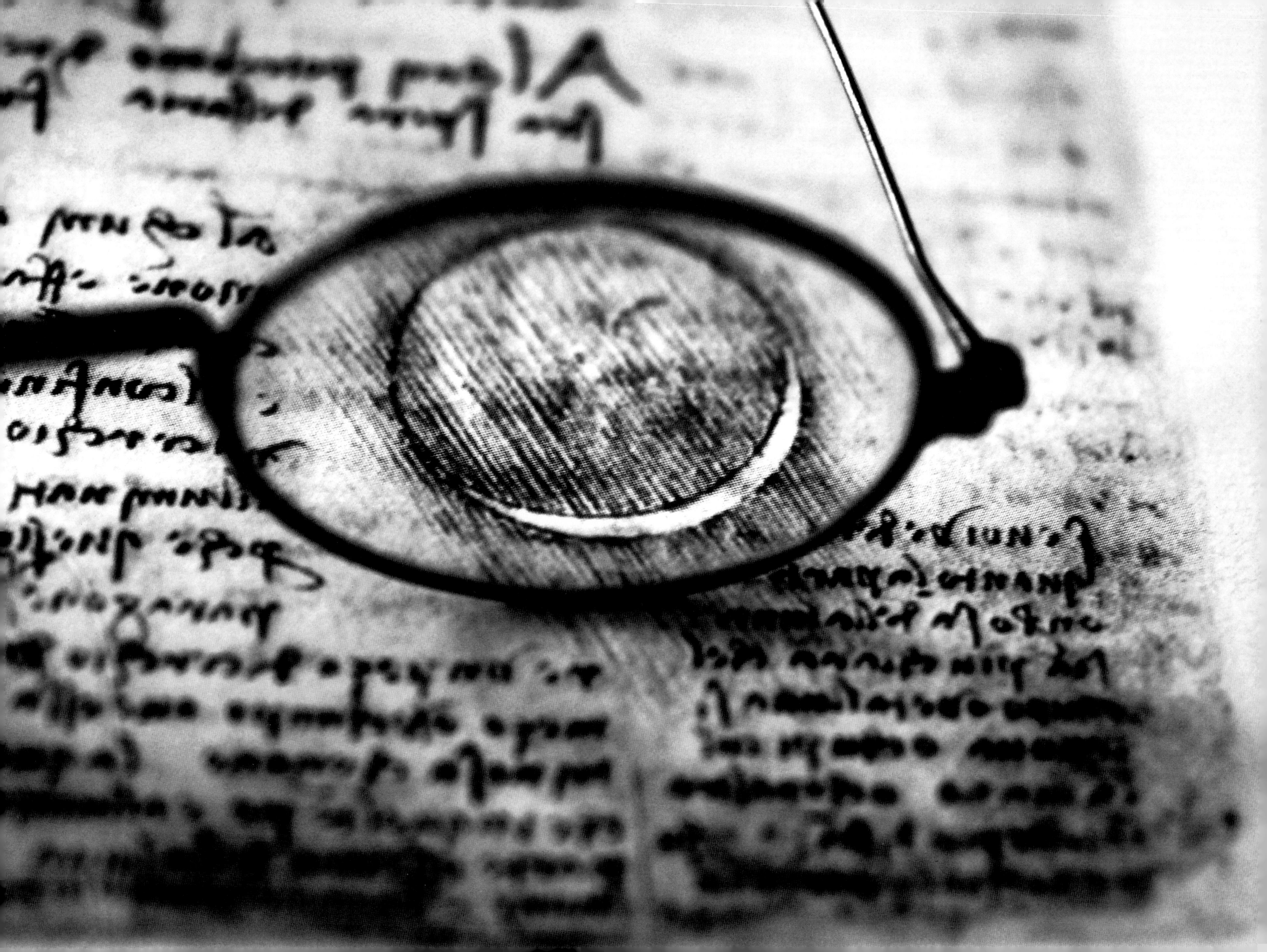

Visceral Visions

Anatomical Study by Leonardo Da Vinci

Royal Library, Windsor Castle

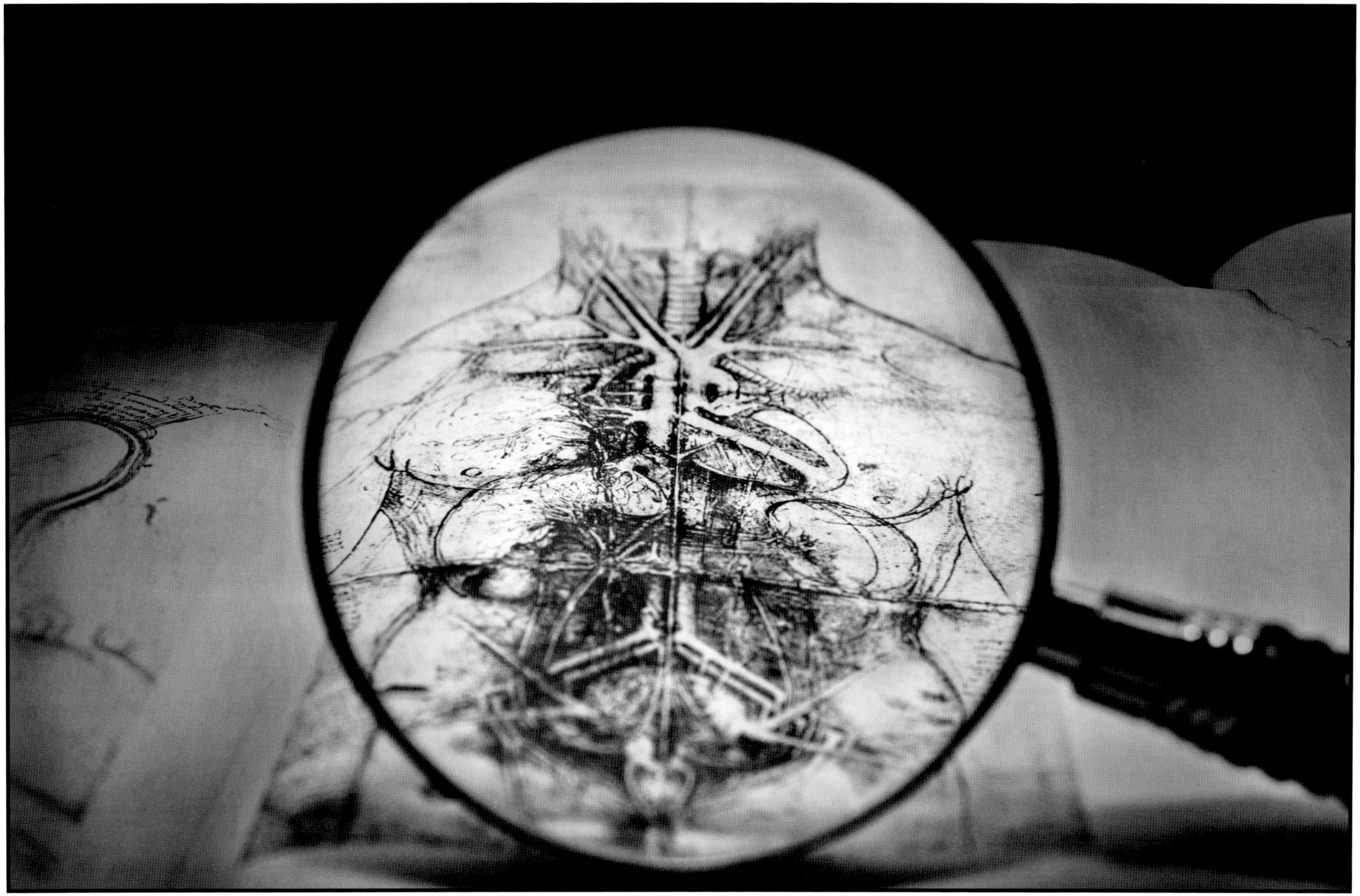

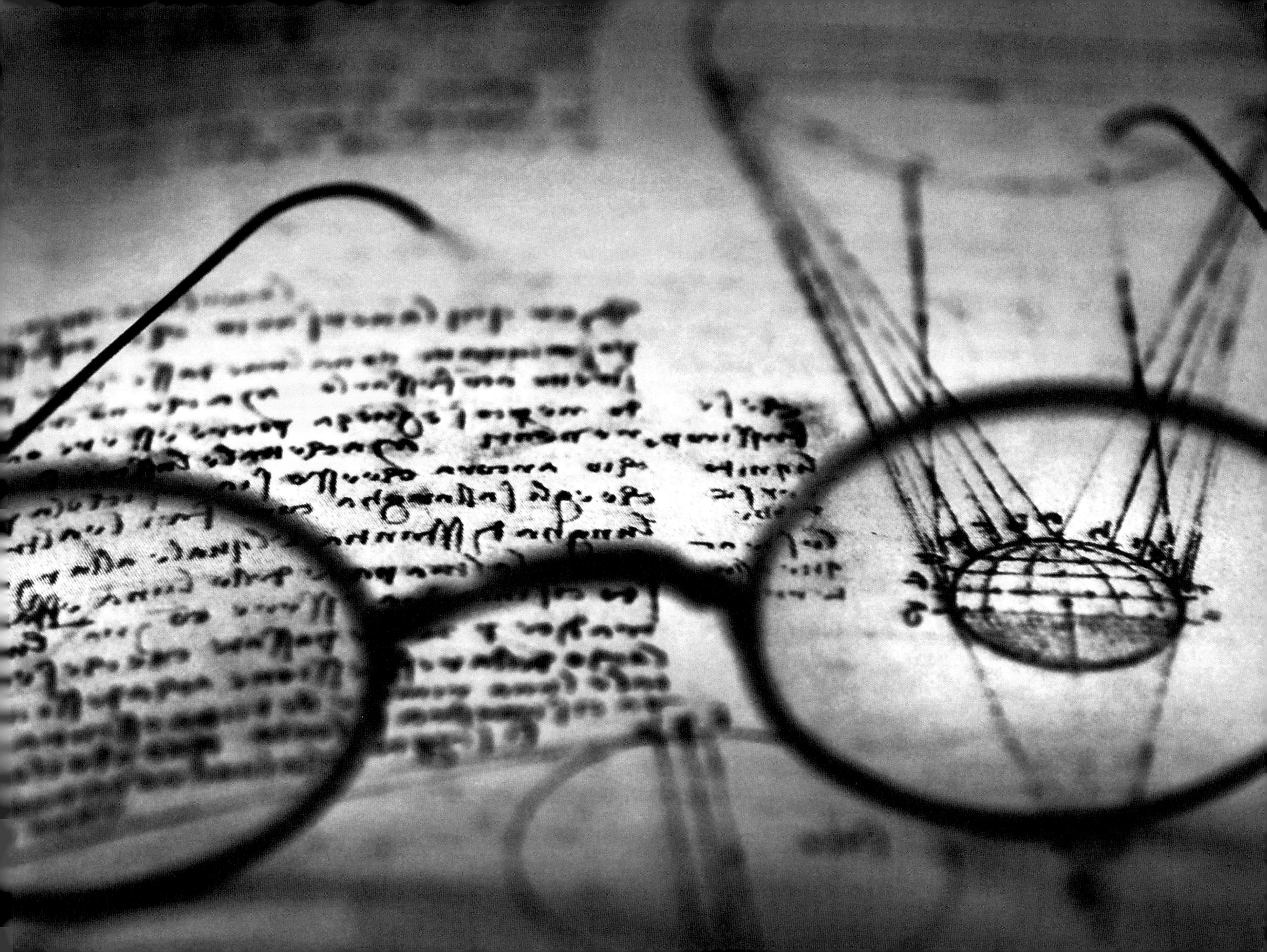

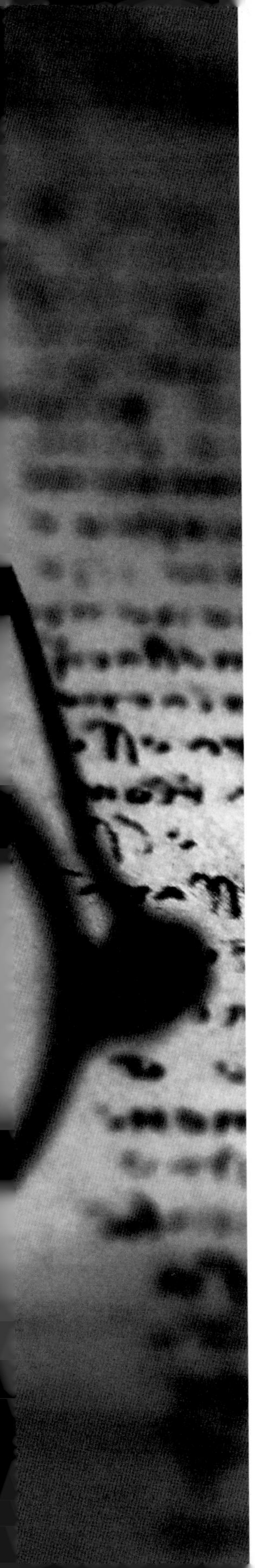

Sun, Earth and Moon

From the *Codex Leicester* by Leonardo Da Vinci, c. 1508-1512

REFLECTING ON A MASTERPIECE

Detail of the *Last Supper* by Leonardo Da Vinci

Milan, Italy c. 1495-1498

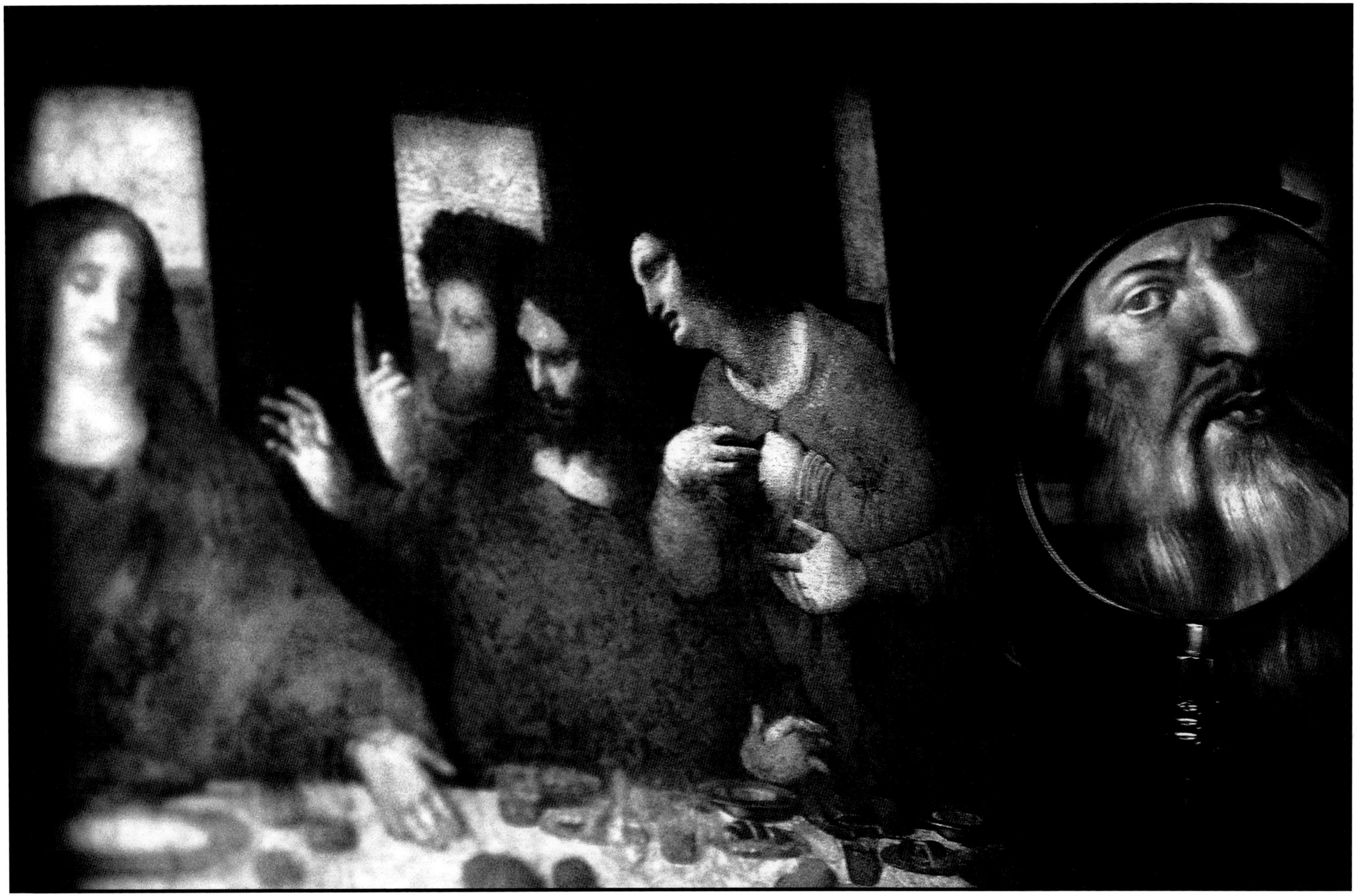

Born Creators

We are all born creators.

Creative brilliance is not a matter of cosmic happenstance.
It is a universally shared truth.

The same genius, which both guided and compelled these great minds in their journeys, lies within each of us if we will but accept it and nurture it.

While the heroes of history each had their own unique gifts and passions, consider for a moment, those traits which they held in common.

To begin with, they were CURIOUS to a fault.

Their lives were built largely around asking intriguing questions. Then once they had discovered an answer, it was put to use as a springboard to ask the next great searching question.

Einstein once attributed his own genius to this very characteristic stating:

"I am no more gifted than any other man, just much more curious."

That same superbly curious man also paid homage to the remarkable power of imagination.

"Imagination is more important than knowledge", he said, "For knowledge is limited to all we now know and understand, while imagination embraces the entire world, and all there ever will be to know and understand."

Indeed, it was largely the boundless IMAGINATIONS of these men and women that allowed them to overcome the gravitational pull of conformity and to see so much further than those around them.

They willfully took wing upon the winds of wonder, letting their minds and hearts soar to dizzying heights. The fear of falling, of dashing great HOPES or impossible dreams did not deter them, because of the incredible perspectives they had gained at such inspired altitudes.

The voices of narrow-minded earth-bound naysayers, with their mantras of "it can't be done" or "that's impossible", could hardly be heard over the melodies of boundless possibility which they had tuned their ears to.

Still, simply imagining and wondering were insufficient; they had to find for themselves the truth behind their ideas, however wild or far-fetched they might seem.

Hence their work was predominantly creative, taking their premonitions to the next level of exploration, invention, or EXPERIMENTATION as they set off in search of answers to questions born of great curiosity.

Those answers often proved elusive and required substantial work, exhausting every possible avenue of inquiry. Eventually though, even the greatest of secrets were revealed to those who would not relent in their pursuit of knowledge and truth.

These were scientists, but no less artists, whose souls reverberated with the stirring compositions of both mind and heart.

They dreamed, they pondered, they explored, and they created in FULFILLMENT of their dreams.

These were discoverers for all ages who lost themselves in learning and who, journeying far beyond the safety of the beaten track, made the epic BREAKTHROUGHS for which the rest of humanity stands in their debt.

Given the magnitude of their lives and accomplishments, it often feels as though the best we "mere mortals" can do is look up to them with profound admiration.

And yet, upon closer examination of these qualities of curiosity, imagination, experimentation, and discovery, one cannot help but note that these are the very traits with which we enter this life with such abundance.

Children never cease to question, they journey constantly through imaginary PLAY, they experiment with everything in reach, and they manage to discover something groundbreaking, in our their minds at least, on a daily basis.

Perhaps one of the great secrets to the success of these historic figures was their ability to see life through the eyes of a child who is never finished discovering, and who can hardly fall asleep at night for the excitement of all that there is yet to do, to learn, and to accomplish.

Perhaps our greatest need as creators is to open ourselves once more to that VISION which we held as young children.

If somewhere along the way we have lost sight of that astounding potential, if the learning voyage has become detached from our life's aspirations, now is the time to return to those dreams we once believed in without reservation.

The tools and technology we have at our disposal would have left our Renaissance counterparts in absolute awe.

Where the ancients had single-room workshops with rudimentary tools, today we can visit a local toy store and purchase science paraphernalia that would have taken them weeks or months to develop. Even the simplest of public libraries has eclipsed their small collections of books, and the Internet has made information access instantaneous. The astrolabe has been replaced by the somewhat less elegant "global positioning system", and complex mechanical models, once painstakingly assembled and tested in a laboratory, can now be built in virtual spaces that simulate their every function.

Hence, our challenge is not whether we can find the information we seek, but whether we will find the INSPIRATION to begin seeking. Our test is to channel our individual passions and gifts in the creation of something extraordinary.

Whatever voyage we choose to embark upon, we should seek to master the secret of the SAIL* which accompanies every great creative endeavor. For when sailing, we learn to harness those same powers which so movingly propelled our predecessors in their own fantastic journeys of discovery.

* Story | Angle | Immersion | Light

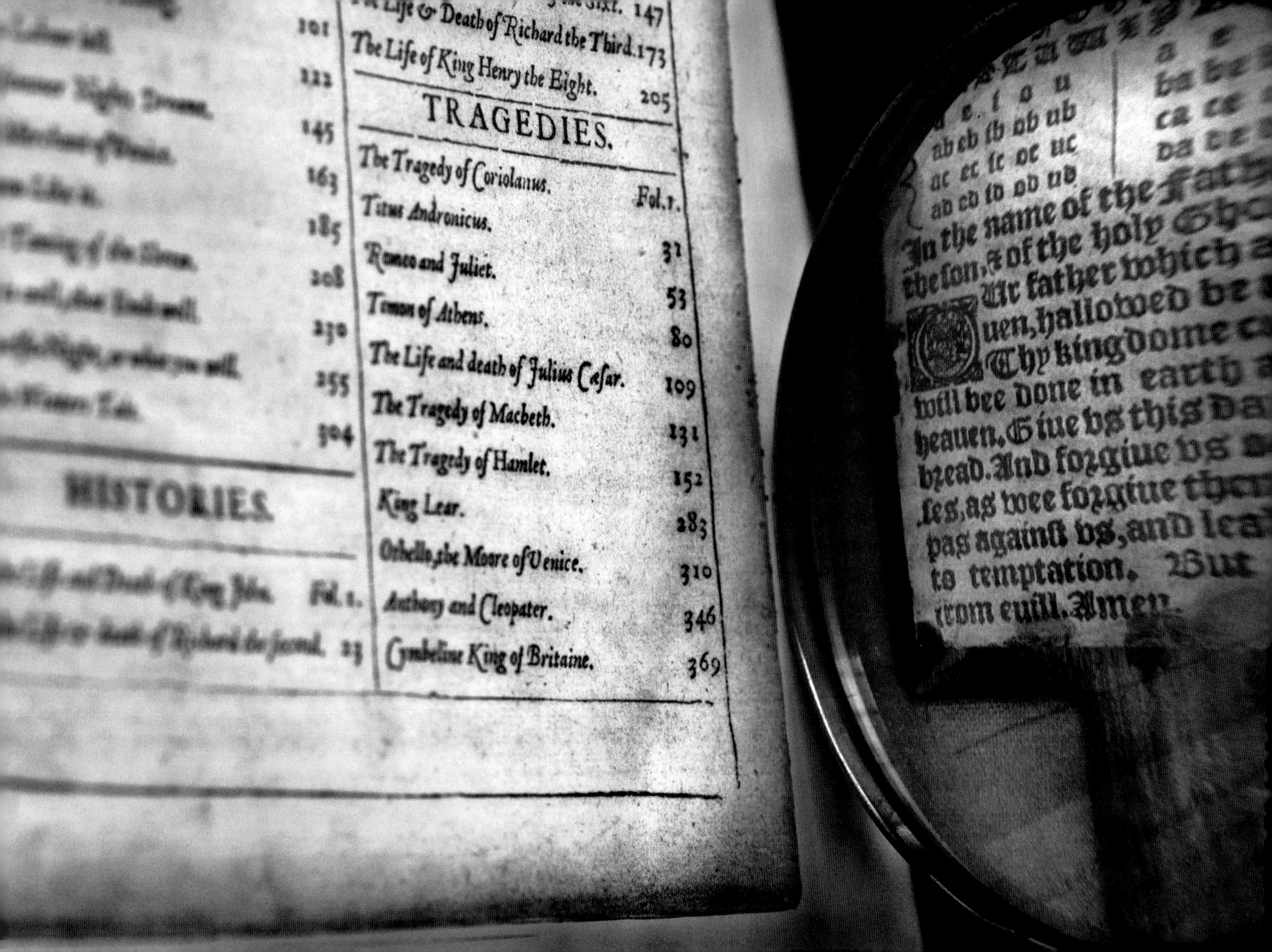
The Life & Death of Richard the Third. 173
The Life of King Henry the Eight. 205
TRAGEDIES.
The Tragedy of Coriolanus. Fol. 1.
Titus Andronicus.
Romeo and Juliet. 31
Timon of Athens. 53
The Life and death of Julius Cesar. 80
The Tragedy of Macbeth. 109
The Tragedy of Hamlet. 131
King Lear. 152
Othello, the Moore of Venice. 283
Anthony and Cleopater. 310
Cymbeline King of Britaine. 346
369
HISTORIES.
In the name of the
the son, & of the holy
Ur father which
uen, hallowed be
Thy kingdome
will bee done in earth
heauen. Giue vs this da
bread. And forgiue vs
ses, as wee forgiue the
pas against vs, and lea
to temptation. But
from euill. Amen.

Story

At the beginning of every creative adventure is a compelling story.

It breathes LIFE into our learning, launching our craft on its way.

Without stories, we are left to drift among the flotsam of disconnected facts and uninspiring data.

Consuming information without an engaging story, is like holding our noses while reluctantly swallowing bitter vegetables, because they're "good for us".

Through a story though, those same vegetables are magically transformed into pumpkin carriages, towering beanstalks, or a pea at the bottom of a hundred mattresses.

Through stories, a whole host of isolated ideas, which are bland on their own, can be mixed into a symphony of tastes and textures to whet our creative palate.

In essence, stories are the real reason WHY we learn.

They kindle our CURIOSITY and ignite exploration's FIRE, compelling us to get up off the couch and to start seeking.

Columbus didn't leave Spain simply because he had nothing better to do on a Friday afternoon than set sail for the edge of the known world.

He cultivated the idea of a new unexplored route until it was burning so intensely within him, he couldn't restrain himself.

He had to set sail. He had to find for himself the passage westward, lest he leave unfinished the story he had begun.

Shakespeare didn't set out to reinvent the English language. He set out to write great stories as only he could.

The words followed as did the fame, but telling the tale was always the aim.

Similarly, we must set out seeking to write our own stories, to give life to our ideas and to find the answers to those questions which will not allow us to rest until we have.

FROM BOY TO BARD

Facsimile Manuscript of an early compilation of Shakespeare's works

Hornbook, London c. 1500 *(Used by children in Shakespeare's day to learn reading)*

Angle

Angles are the AERODYNAMICS of exploration.

They give us the creative capacity to look at life from various views. Without this gift, we are bound to an unchanging or at least a very narrow view of the world.

As we seek out such angles though, our vessel is lifted from the sea to sail among the clouds, and from thence to the stars and beyond.

Nothing opens our eyes like a newfound perspective on the vast oceans of knowledge yet unexplored.

Creative angles can also lead us cultivate a deeper sense of intellectual HUMILITY and AWE at how much we have yet to learn.

They provide the opportunity to approach the world with the new eyes of a child each day.

Seeking meaningful angles encourages deep observation and the contemplation of numerous possibilities, such that we are less likely to take the details of life for granted.

We learn to look more closely, to consider more fully and so doing, we begin to truly see.

We see the world, we see others and ultimately we come to see the truth and the potential within our own selves; entirely free of the restrictions we had once assumed to be insurmountable as our dreams fall within our reach.

Galileo's own world was one of closed minds and prescribed perspectives. The very acts of thinking and exploration had very strict rules as to how one could proceed.

As he boldly sought out his own path and understanding, Galileo came to see the world for what it truly was: a single point sailing through the vastness of space.

The heavens were laid open to his view, and his soul SOARED among the stars.

As we spread the wings of our own wide-angle perspectives, we too will uncover the inspiration to RISE above the skeptics and the cynics, carried by the currents of CREATIVE POSSIBILITY.

Micrographia Title Page by Robert Hooke, 1665

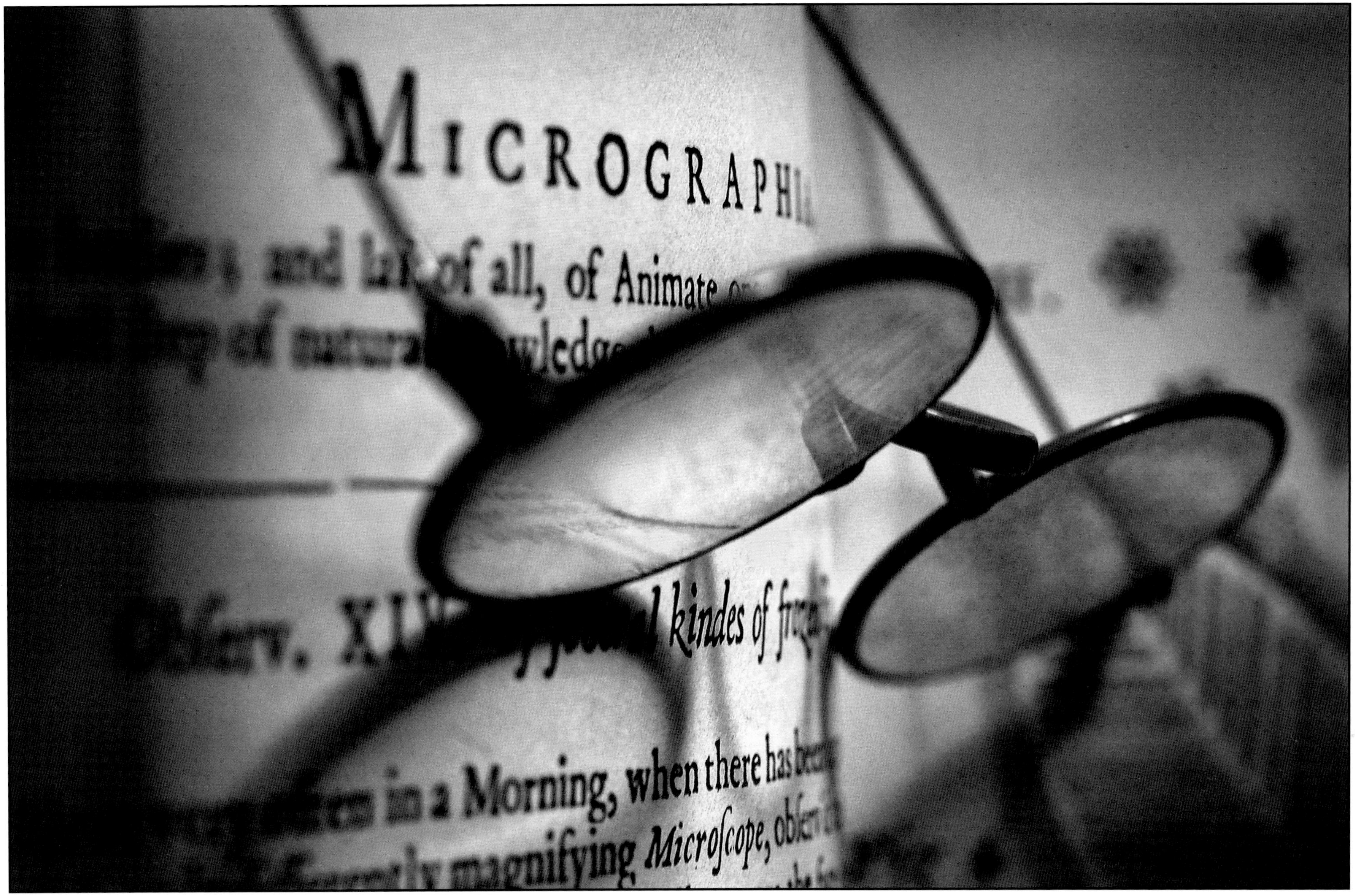
MICROGRAPHIA
and last of all, of Animate
kindes of
in a Morning, when there has
magnifying Microscope,

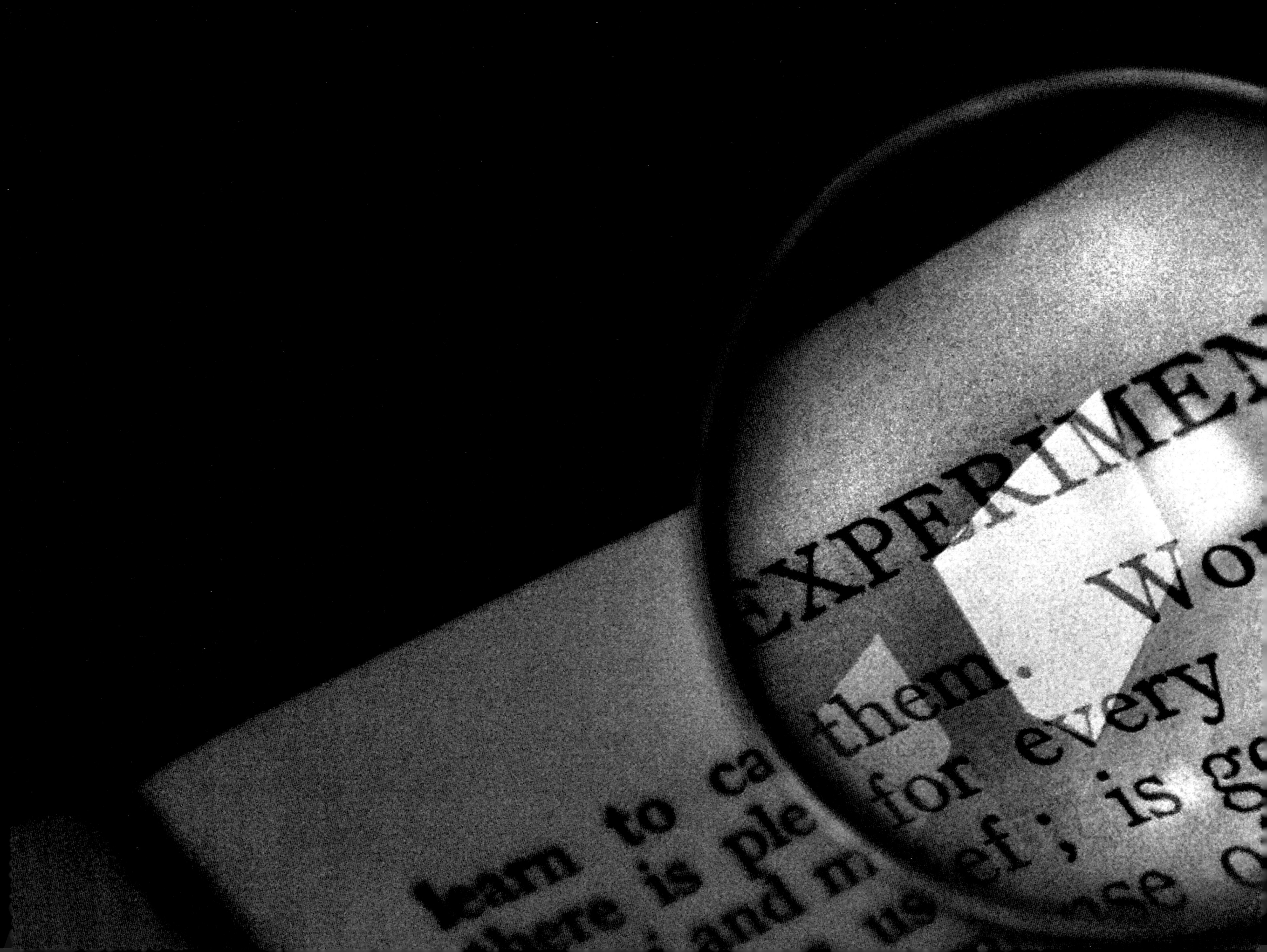
EXPERIMEN
Wo
them.
for every
learn to ca
there is ple

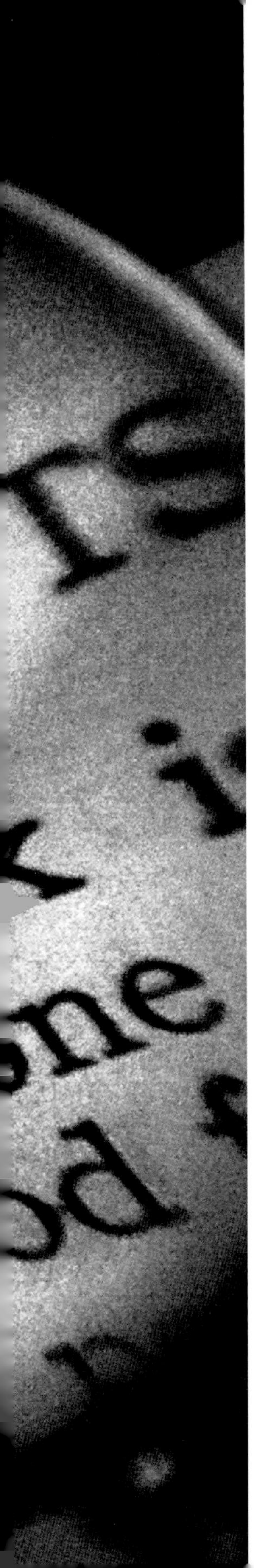

IMMERSION

With a dream woven into a compelling story and guided by an innovative angle, our next step is the creative PLUNGE.

Until we have immersed ourselves, our learning amounts to little more than a superficial skimming of the sea's surface.

We dip our toes in here and there to test the waters, sometimes shuddering at the cold, but never discovering the world beneath the waves.

Immersion entails DIVING DEEPLY, for the sea's greatest treasures lie at her very depths.

Beneath the commotion of the surface, there exists a realm of serenity where time stands still and inspiration is born.

When immersed, our understanding is quickened and we learn to lose ourselves in the MAGIC OF THE MOMENT.

Our petty cares depart with the rising bubbles as the immense ocean of knowledge before us coaxes ever more profound questions, poignant REFLECTION, and inventive experimentation.

Each experience and every piece of knowledge gained is put to use as we probe more deeply, seeking to uncover those PEARLS of pure insight and inspiration.

True, not every oyster will contain such treasure, but there is no way of telling until we have come close enough to explore.

Immersion calls for action and diligence if we are to realize our dreams. It is the labor of learning. Yet, when immersed, such work is never laborious.

So it was with the great discoverers of history for whom learning was never merely a matter of information collection

It was proof of their being alive!

They were SEEKERS and WONDERERS, wanting to KNOW great things because they wished to DO great things with that knowledge.

Losing ourselves as they did, we come to focus our learning on our own DESIRES to create, to fulfill, and to become something more today than we were the day before.

LIGHT

Our ultimate hope, in all this questing for knowledge, wisdom, and truth, is that it will lead us to the EXHILARATION of uncovering some of the answers.

The spark that links all the searching and seeking to that EUREKA! epiphany of enlightened understanding is difficult to capture in words.

It is the creative muse which courses through the veins, opening wide the floodgates of inspiration and beckoning artist, scientist, and discoverer alike to devote their every faculty to the fulfillment of their work.

It is what Handel felt as he composed for days on end, lacking even the desire to eat, while through him came that timeless masterpiece of music, an EXALTING canticle to the King of Kings.

It is the PASSION that blazed so intensely in the bosom of Michelangelo as his brush adorned the ceiling of the Sistine Chapel until his own fingers touched there where God's and Adam's fingers meet in a sublime symbol of the union between the CREATOR and his creation.

At its most basic form it is light...the great revealer of mysteries and the fire of the imagination.

At its transformational touch, even the most basic and rudimentary subjects are endowed with a sense of the EXTRAORDINARY and the enchanting.

It is the sustenance to every seed of study and the reassurance that learning's labors will ultimately bear fruit.

The sweet taste of that fruit is also difficult to describe, but it is so SATISFYING that, having once eaten, the seeker never waits long before pursuing such light in earnest once more.

Like the wind, it is impossible to command, but an awe-inspiring force when harnessed.

While we cannot force inspiration, we can hoist the creative sails of our minds such that, when inspiration lights upon us, we are prepared to capture a portion of its guiding, driving rays, to venture boldy into the great unknown.

PRISM LIGHT

Title Page from Opticks, Sir Isaac Newton, 1704

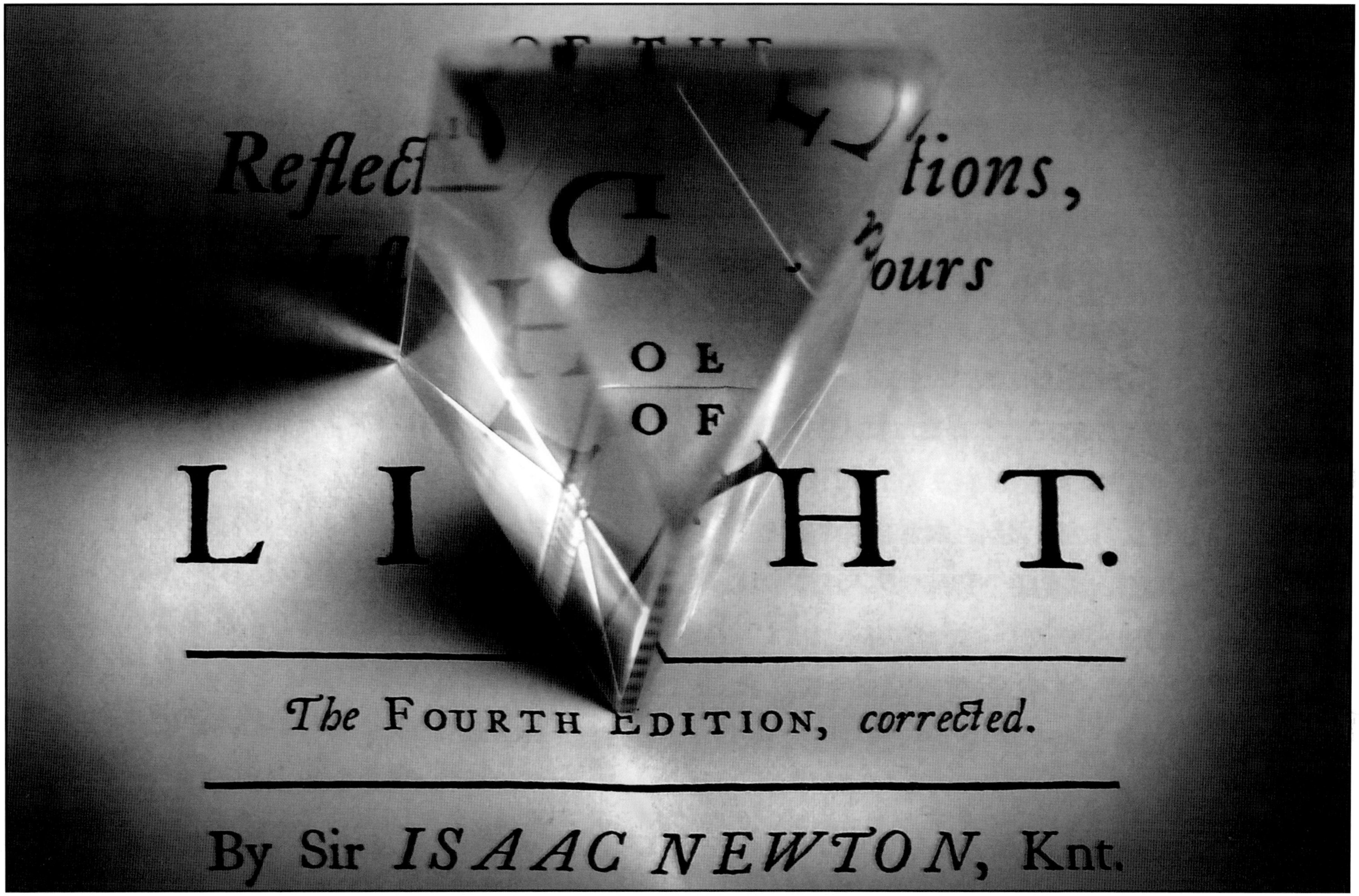
Reflec tions,
ours
OF
L I H T.
The FOURTH EDITION, corrected.
By Sir ISAAC NEWTON, Knt.

Creator and Creation

Detail of the Sistine Chapel by Michelangelo Buonarotti

Rome, Italy c. 1508-1512

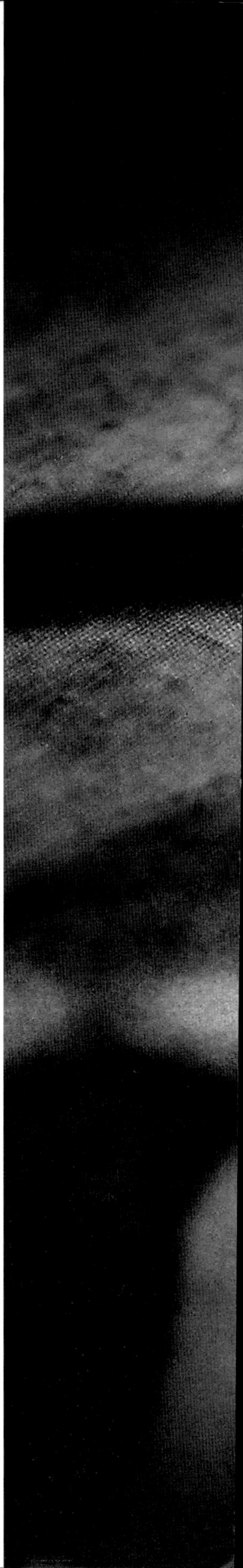

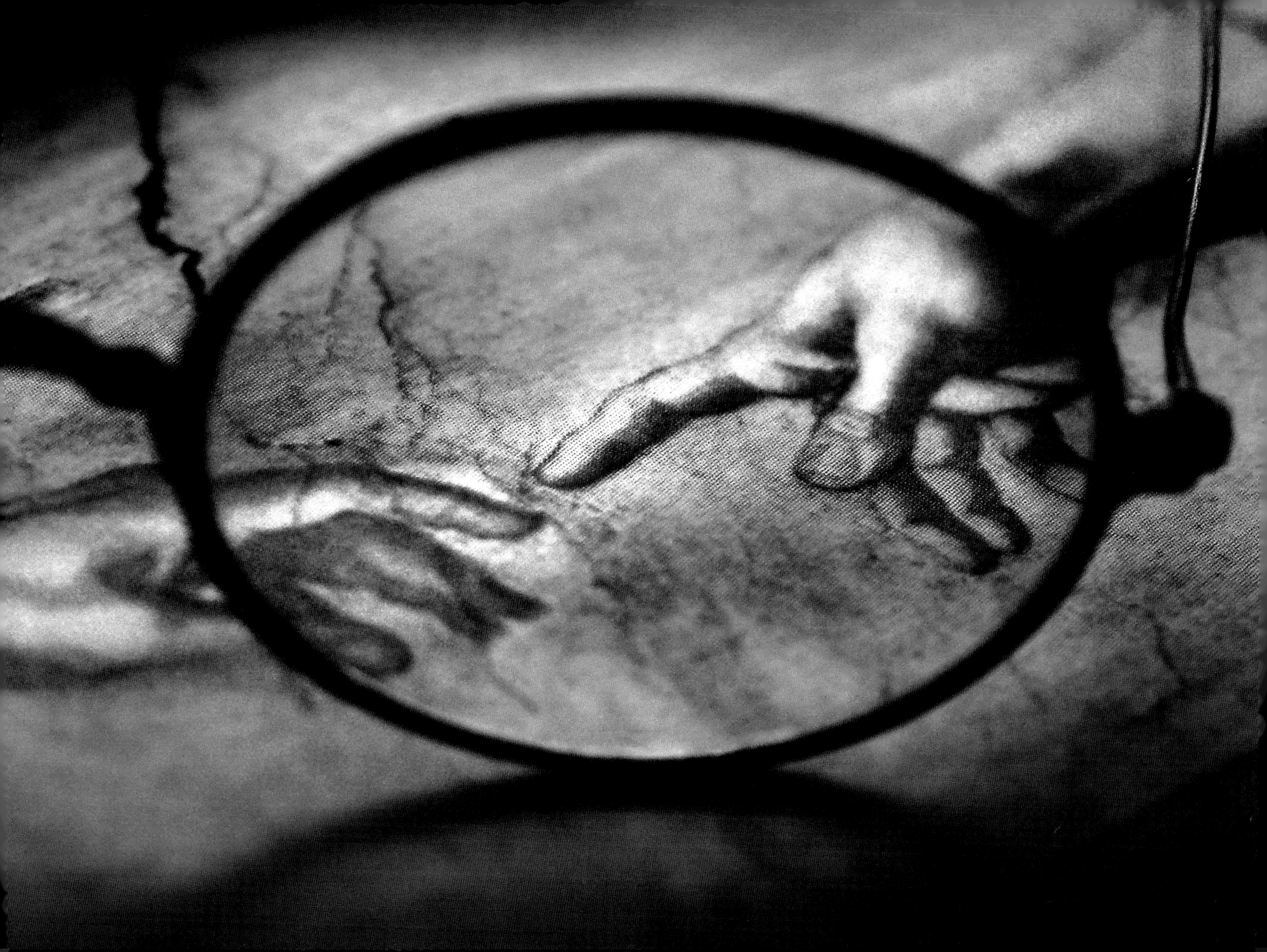

The Composer

From Handel's *Messiah*, 1741

(The Original Manuscripts in Facsimile)

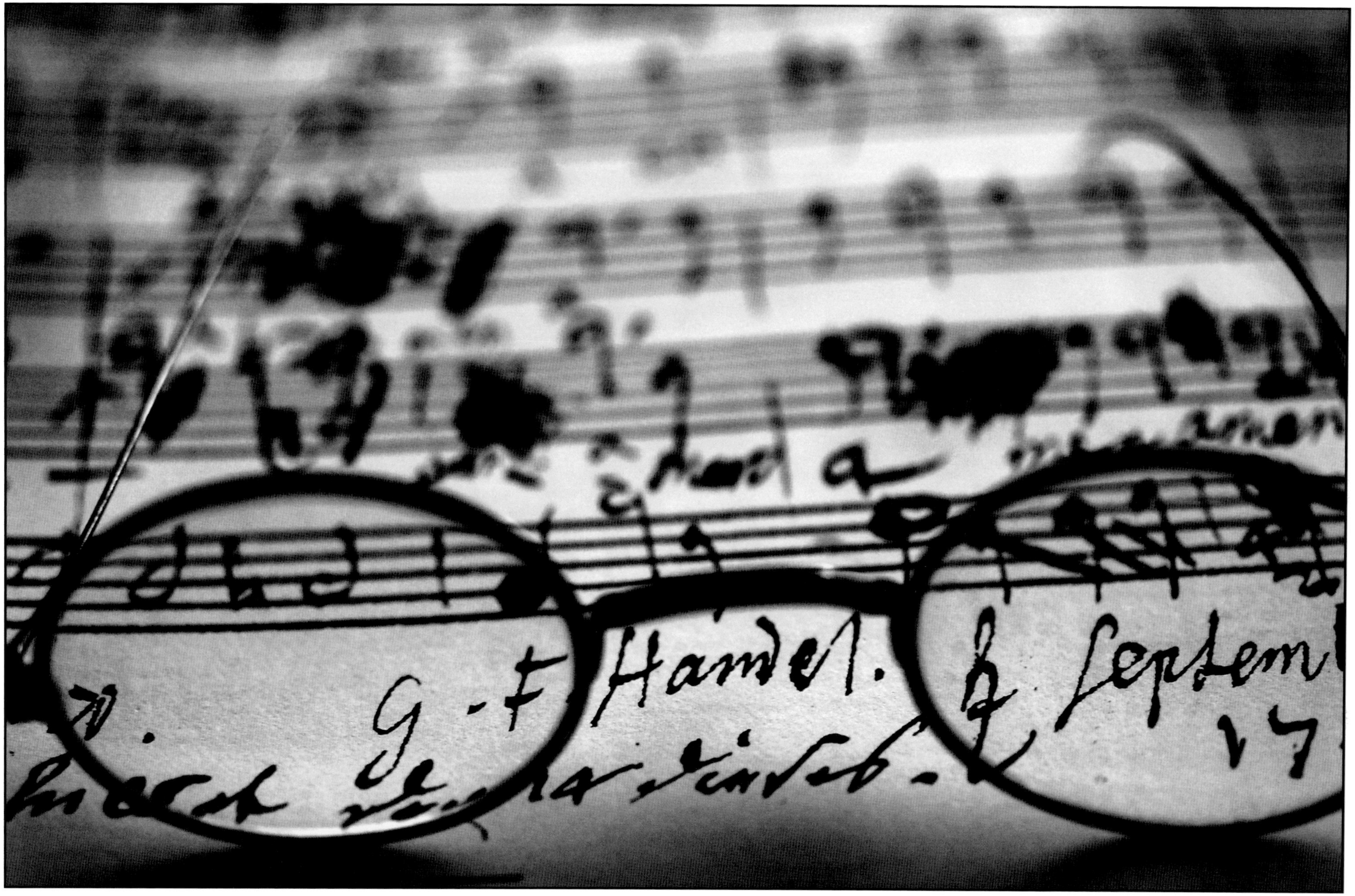
G. F. Handel.

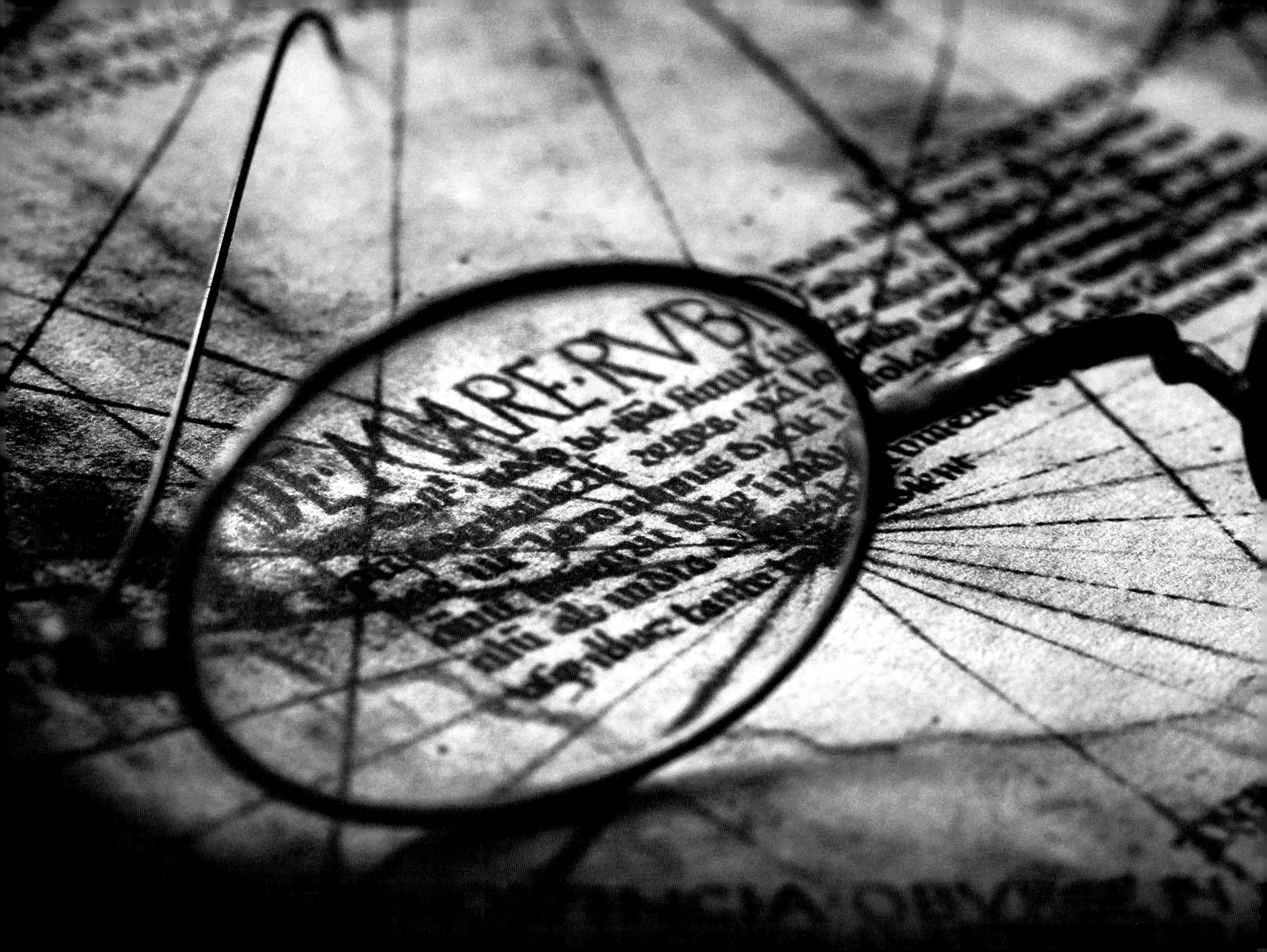
MARE·RVB

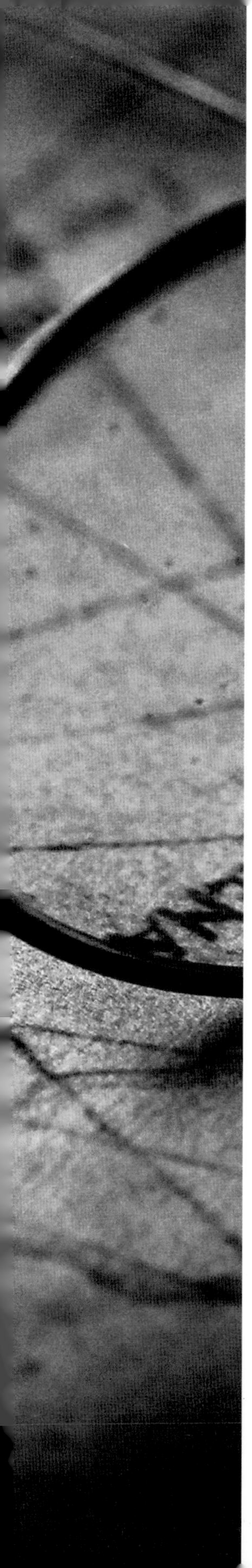

MERIDIANS

From Imago Mundi by Pierre d'Ailly, 1410

SUNDIAL

Universal Horizontal Sundial, Johannes Kooch, Stockholm, Sweden c. 1650 - 1679

Courtesy of the History of Science Department, Collection of Historical Scientific Instruments, Harvard University

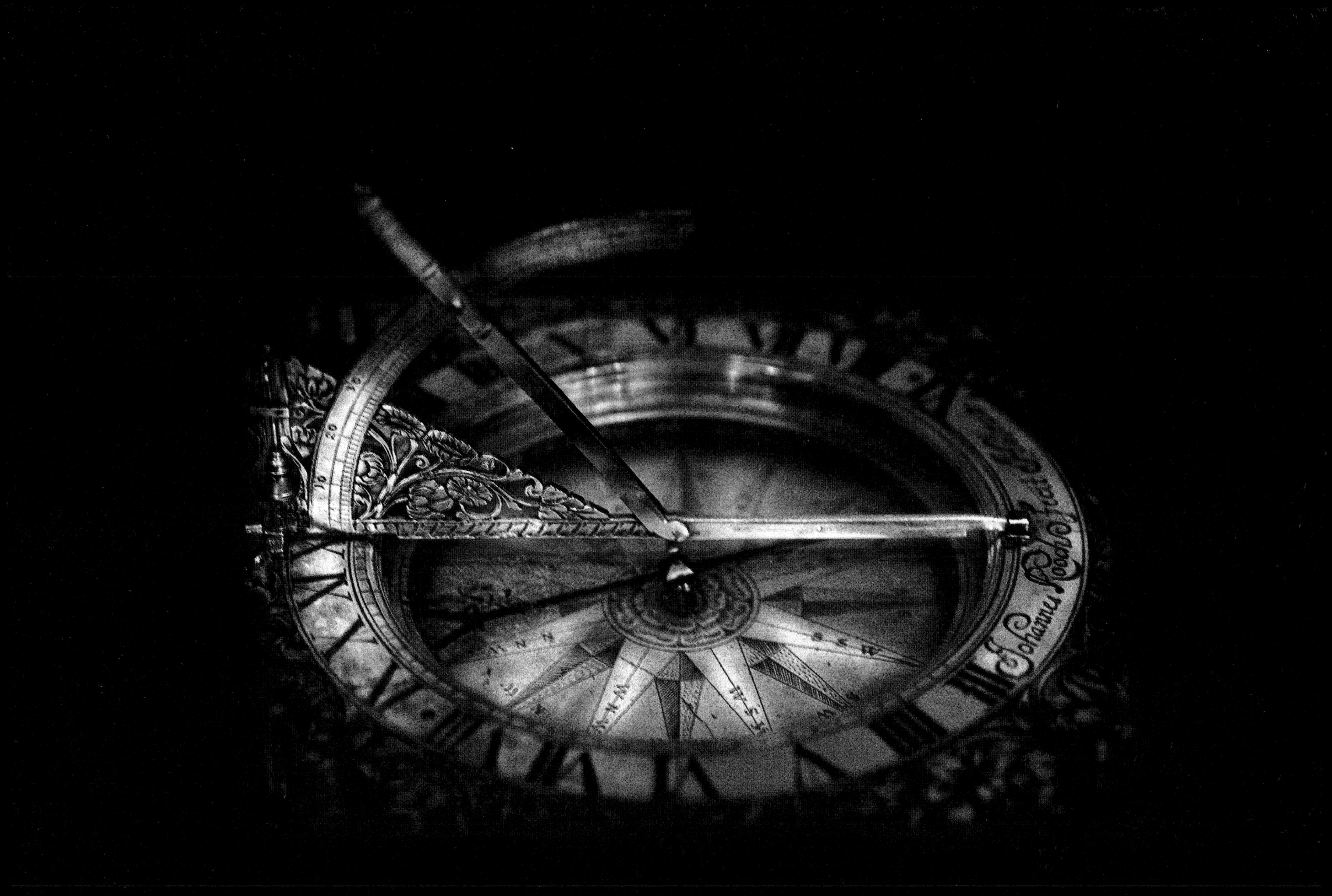

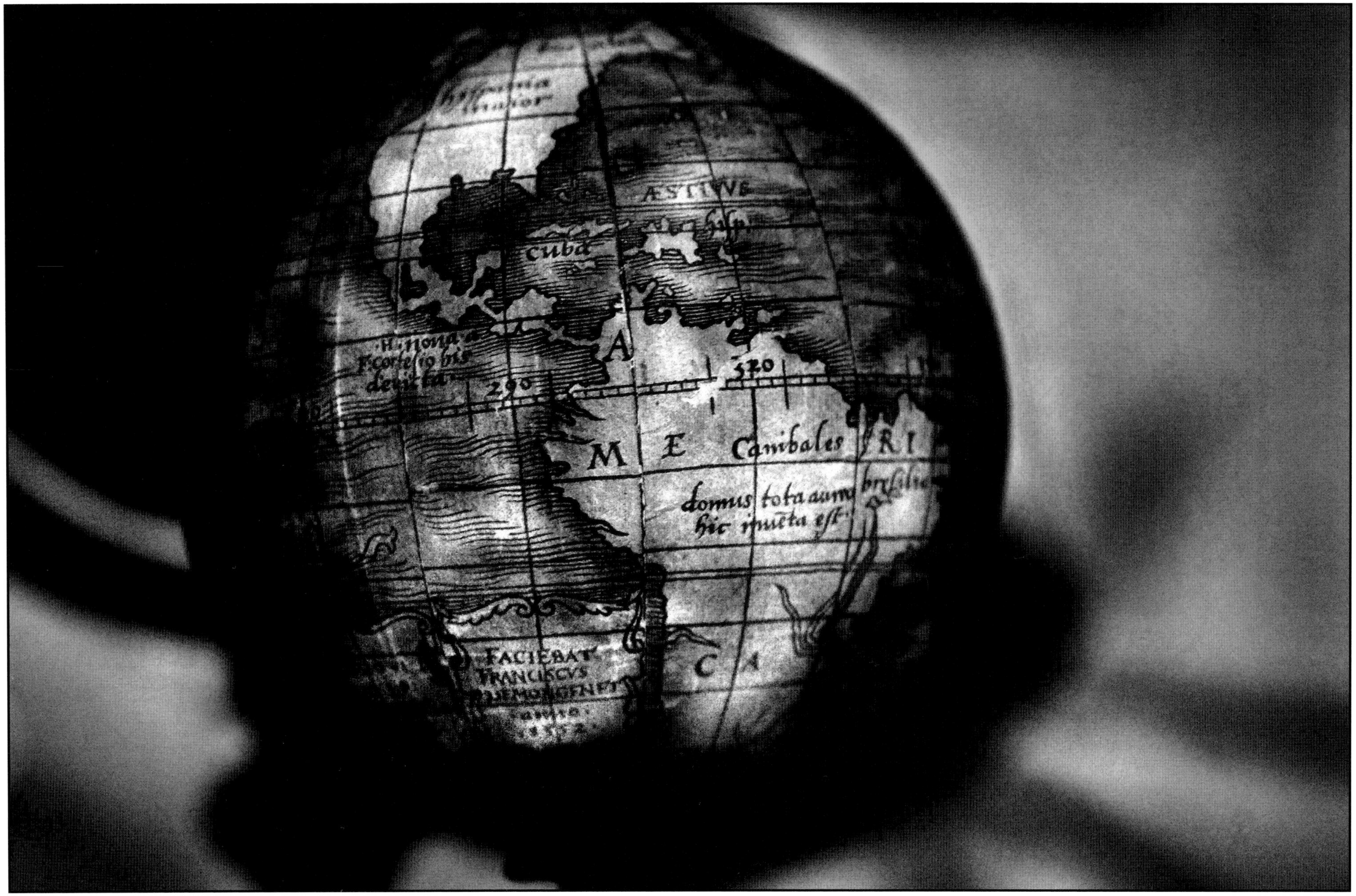

cuba
290
320
Canibales
FACIEBAT
FRANCISCVS

Terrestrial Globe

Francois Demongenet, Vesoul, France, 1552

Courtesy of the History of Science Department, Collection of Historical Scientific Instruments, Harvard University

HALLELUJAH

From Handel's *Messiah*, 1741

(The Original Manuscripts in Facsimile)

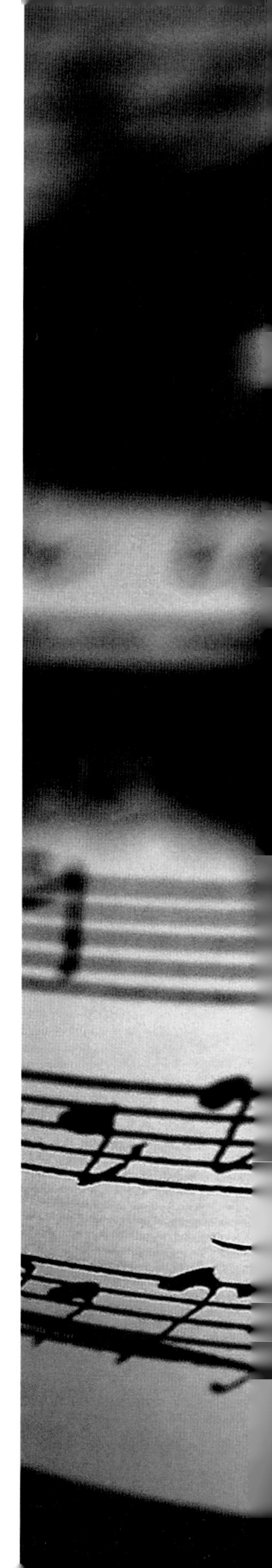

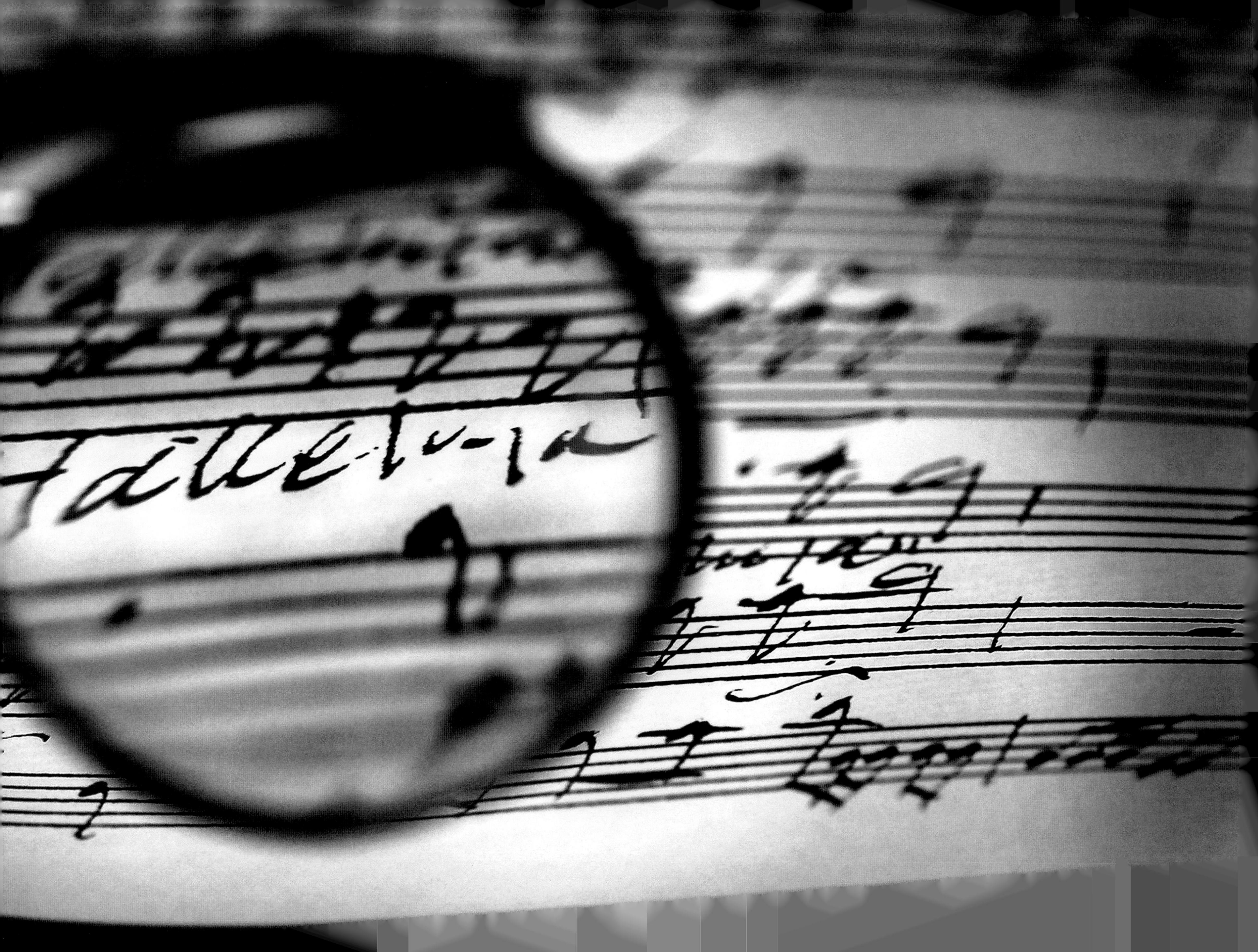

Conclusion

Losing ourselves in anything is a tall order.

So much of our lives are built around seeking a sense of control.

We cleave to comfort, and we tend to avoid those less-trodden paths whose ends are uncertain.

Calls to FORSAKE COMFORT and to open ourselves to emotional or visionary courses are often doused with a cold splash of skepticism.

Yet in a day and age where information moves at light speed and technology changes in a heartbeat, there is no better starting place for seeking success than in cultivating our individual senses of curiosity and creativity.

Yes, losing ourselves requires a tremendous degree of TRUST. We push ourselves beyond the rational realms of comfort and control, so that we may take part in something greater than self.

In such a state, we cherish the knowledge we have received and we come to enjoy the journey of seeking.

We act upon it, build upon it and still we remain OPEN to whatever further TRUTHS lay yet undiscovered.

Never can we say that we have gained enough knowledge, nor can we allow ourselves to rest on our laurels.

No matter the question answered, the test passed or the degree adorning our wall, there will always be more to learn—INFINITELY more.

So, while epiphany and emotional pinnacles mark the culmination of every great educational endeavor, they in no way imply a final destination.

For with every bit of light we receive and every mystery we uncover, we have done little more than open new paths to even greater vistas and even deeper seas.

Every great answer will likely lead to a series of even more profound questions, and so the journey and the adventure continues…ever seeking…ever finding…ever…

Lost in Learning

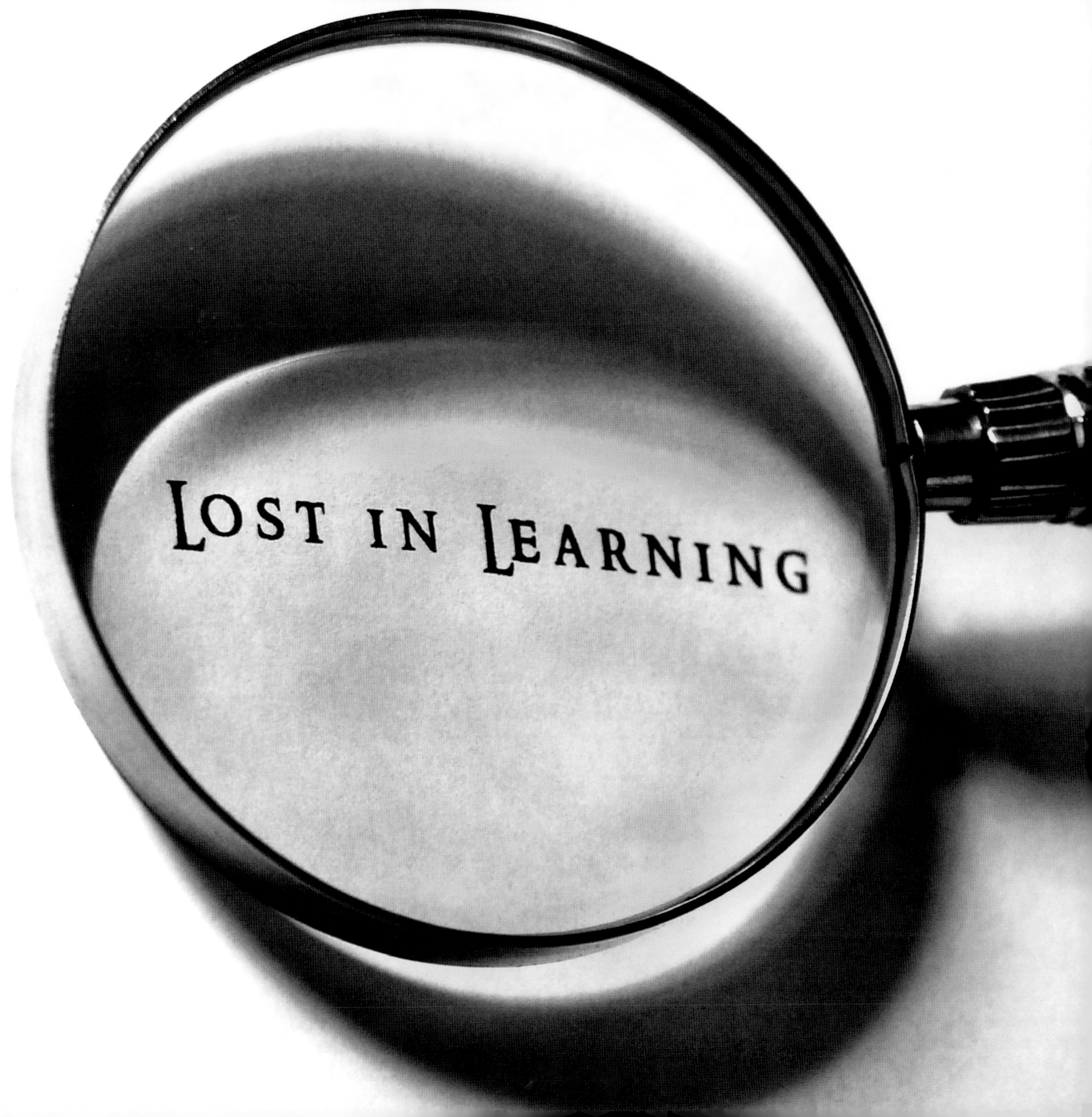
LOST IN LEARNING

Athenaeum Publishing
P.O. Box 1381
Newburyport, MA 01950
857/272-6318.

Printed in China

Publisher's Cataloging in Publication

Timothy, Eva Koleva.
Lost in learning : the art of discovery / Eva Koleva Timothy ; foreword by Ralph Windle ; essay by Adam Timothy.
p. cm.

ISBN 978-0-615-34394-5

1. Philosophy–Pictorial works. 2. Education–Philosophy. 3. Education–Pictorial works. 4. Philosophy–History. I. Windle, Ra lph. II. Timothy, Adam. III. Title.

B51.8 T56 2010
109–dc22 2010921396

ATTN: Quantity discounts are available to your company, educational institution or organization for reselling, educational purposes, subscription incentives, gifts, or fundraising campaigns. For more information please contact the publisher at: info@athenaeumpublishing.com.

Book Design Editor
Elizabeth Avedon

Content Editors
Richard Eyre & Adam Groff

A special thanks to Sara Schechner and Martha Richardson of Harvard's History of Science Department for access to the University's collection and for their insights into those instruments which facilitated the explorations of an age.

Exhibitions and Prints

Lost in Learning is a traveling exhibit designed for Universities, galleries, museums, science centers and other educationally focused institutions. For further information contact: exhibits@illumea.com

Limited edition prints of the Lost in Learning collection are available through Illumea Photography contact: artsales@illumea.com / 617-470-4789.

www.lostinlearning.com